Molding of Thermosetting Plastics

W0071804

Tony Whelan and John Goff

VAN NOSTRAND REINHOLD
New York

Library of Congress Catalog Card Number
ISBN 978-0-442-30318-1   ISBN 978-1-4615-9759-9 (eBook)
DOI 10.1007/978-1-4615-9759-9

First published in the U.S.A. in 1990 by
Van Nostrand Reinhold
115 Fifth Avenue
New York, New York 10003

Van Nostrand Reinhold International Company Limited
11 New Fetter Lane
London EC4P 4EE, England

Van Nostrand Reinhold
480 La Trobe Street
Melbourne, Victoria 3000, Australia

Nelson Canada
1120 Birchmount road
Scarborough, Ontario M1K 5G4, Canada

16 15 14 13 12 11 10 9 8 7 6 5 4 3 2 1

Library of Congress Cataloging-in-Publication Data
Whelan, Tony.
     Molding of thermosetting plastics/by Tony Whelan and John Goff.
       p.   cm.
     ISBN 978-0-442-30318-1
     1. Thermosetting plastics.   2. Plastics – Molding. I. Goff, John.   II. Title.
TP1180.T55W46   1990   89-78425
668.4'22–dc20   CIP

PREFACE

During the years 1987 and 1988 we published a series of articles on the molding of thermoplastics materials in the magazine British Plastics and Rubber (B P & R). These articles were very well received and we also received a large number of requests for reprints.

In order to cater for what is obviously a need in the thermoplastics molding industry, we therefore brought the information together and produced it in the form of a book.

We can only hope that it serves you well and that you find the information useful. We in turn would like to thank the editor of the magazine B P & R for helping us in this matter.

Thanks are also due to our many friends and colleagues throughout the molding industry for their useful help and advice, in particular the company Moldflow (Europe) Limited deserve a special mention as they allowed us to extract information from their extensive data base.

The information presented in this publication was based on our current knowledge and experience and represents the opinions of us, the authors, and not necessarily the opinions of any organization. Such information is given in good faith but in view of the many factors that affect processing and properties, the data so presented does not relieve processors from the responsibility of carrying out their own tests and experiments: neither does the data presented imply any legally binding assurance of certain properties, ease of processing or of suitability for a given purpose. It is also not our responsibility to ensure that proprietary rights are not infringed and that relevant legislation is observed.

John Goff and Tony Whelan

Contents

PLASTICS AND POLYMERS

The polymers which are processed by melt processes, such as molding and extrusion, may be divided in various ways, for example, they may be divided into thermoplastics, thermosetting plastics and, into elastomers. Each type of material may, in turn, be extensively modified by, for example, the use of fillers.

THERMOPLASTIC MATERIALS

A thermoplastics material is one which softens on heating and hardens on cooling: this heating and cooling process can be repeated a large number of times. Approximately 80% of all plastics used throughout the world are thermoplastics: this important group of materials has been divided into four groups:-

    i)   commodity or bulk thermoplastics;
    ii)   engineering thermoplastics; and,
    iii)   thermoplastic elastomers or rubbers, and,
    iv)   blends or alloys.

At one time everybody seemed to know what these terms meant but now, because of materials modification, the position is far less clear with, for example, plastics such as PP (see Table 1 and Table 2 for a list of abbreviations and, for letters used to modify abbreviations) sometimes included in a definition of engineering plastics and sometimes not. Even this simple example illustrates a very important point: it is possible to lift, or move, a plastics material from one category to another. In the case of PP, a commodity plastics material, it can be changed into an engineering plastics material by materials enhancement.

The practice of materials modification is as old as the polymer industry itself, For example, sulfur is added to cross-link rubber; rubber is added to polystyrene (PS) and, the effect is remarkable as it changes PS from a material with poor impact strength to one which is good enough for the manufacture of stiletto shoe heels. In both cases there is not a simple addition; one material is chemically grafted, or bonded, to another. In the case of the second example however, the new polymer (high impact polystyrene, HIPS or, TPS or, IPS) can be reprocessed as it is a thermoplastics material. Cross-linked rubber cannot be easily reprocessed as the chemical links are not thermally labile i.e. not capable of being broken on heating. What is required is a material which can be processed like a thermoplastic and yet, has many of the properties of a cross-linked plastics material: this is one reason for the current interest in polymer alloying.

COMMODITY OR BULK MATERIALS

This group of materials covers the major materials such as polypropylene, polyvinyl chloride, polystyrene and polyethylene (PP, PVC, PS and PE). Each of these divisions covers a number of sub-divisions, for example PP covers homopolymers and copolymers, PVC covers rigid and plasticized, PS covers general purpose PS and toughened PS (TPS, or IPS, or HIPS) and PE covers LDPE, LLDPE and HDPE. The importance of bulk thermoplastics cannot be over-emphasized; approximately 70% of all plastics used fall into this category.

These materials are used so extensively because the raw materials on which they are based are relatively cheap and, conversion into plastics which have useful properties is fairly easy. It is also now being realized that the properties of this class of materials can be up-graded so that they compete with other more exotic, and therefore more expensive, plastics. A case in point is UPVC; because of the inherent flame resistance if this material, it has been developed so that it can now compete with engineering thermoplastics in some applications.

ENGINEERING THERMOPLASTICS

Engineering plastics are a group of polymers which offer a combination of some of the following, high strength, stiffness, toughness and, resistance to wear, chemical attack and heat.

Any discussion of engineering thermoplastics must concentrate on materials such as polyamides, acetals, polycarbonates, thermoplastic polyesters and modified polyphenylene oxide (also known as modified polyphenylene ether or PPE). Greater than 90% of all the engineering plastics used in the world, fall into one of these categories (approximately 8% of all plastics usage is accounted for by these five categories). Other materials such as the sulfones and the modified polyimides only account for a very small percentage (approximately 1 to 2%) of the total usage of engineering plastics. This relatively low consumption occurs, because most of the engineering plastics use, is in metal replacement where exotic property (and therefore expensive property) characteristics, are not required. The 'tonnage', engineering thermoplastics are used in a wide range of applications even though they have comparatively low heat resistance (below 150°C or 302°F) and relatively low creep resistance especially at high, or elevated, temperatures.

THERMOPLASTIC ELASTOMERS

Those materials which are referred to as 'thermoplastic elastomers' (TPE) or, as 'thermoplastic rubbers' (TPR) are a comparatively new group of materials but already they have become significant in terms of both tonnage and in terms of value. This is because they are creating new markets as well as replacing both plastics and traditional rubbers in some of their applications. They exhibit some of the properties associated with traditional, or vulcanized, rubbers, but they also exhibit the speed, and ease, of processing of thermoplastics; the ability to be re-processed is also important. Of major importance is the wide color range possible from this type of material. TPEs may be based on specifically manufactured polymers (e.g. styrene butadiene styrene or polyether ester elastomer - SBS and PEEL) or, they may be based on blends of rubbers and plastics. For example, a widely used material is that obtained by mixing PP with an ethylene propylene rubber. A two phase material results in which there is a plastics phase and a rubber phase. In the simplest cases, the rubber is not crosslinked or vulcanized but, is simply dispersed throughout the plastics material. There are other materials which also contain a dispersed rubber phase but, in this case, the rubber is partially vulcanized: these may be known as thermoplastics vulcanizates (TPV).

It is believed that the usage of TPEs will increase to such an extent that they will displace traditional rubbers from many applications. Such traditional elastomers/rubbers will continue to be used for applications requiring high temperature resistance and dynamic performance but TPEs will serve better for all other applications.

The reasons cited for the anticipated decline of traditional rubbers, in favor of TPEs, include the availability of a variety of TPEs of consistent quality, the elimination of customer formulation and compounding, easy and rapid processing, elimination of non-reusable scrap and, the often superior resistance of TPEs to chemical attack: it is often possible to replace a range of rubber mixes by one TPE because of the superior chemical resistance of the TPE.

BLENDS

There are a number of reasons for blending materials and these include the generation of unusual materials, extending the performance of existing plastics, extending this performance quickly, extending the performance cheaply and, re-using plastics materials. Both time and cost are against the development of new plastics. It may, for example, take fifteen years to develop a new plastic and cost many millions of dollars: a new blend can be produced for a fraction of this cost. The other factor is that new polymers, as envisaged by the manufacturers, have rarely satisfied the market for very long and, once original markets have developed, demands usually arise for performance improvements.

The blends which result are often produced on twin screw

extruders using what is called reaction compounding. As many twin screw machines are modular in design, their construction can be tailored to suit a particular resin blend thus facilitating production and giving the all-important, blend consistency. Because of the relatively poor performance of many engineering plastics in some applications, and/or because of their high price, recent attention has focussed on engineering polymer blends (EPB).

Engineering polymer blends (EPB) can be roughly divided into blends of an engineering plastic or resin (ER) with a commodity plastic, blends of an engineering plastic with another engineering plastic, blends of an engineering plastic with an elastomer and, blends which contain three or more polymers. We can therefore get combinations such as PPO/PS; PPO/PA; PC/ABS; PET/PBT: PBT/PC/SAN etc.; each of the blends may in turn be filled. When blends are made the objective is to simultaneously optimize blend formulations, in respect of several properties important for a particular application, sacrificing those which are not important.

THERMOSETTING PLASTICS

A thermosetting plastic, or a thermoset, may be defined as a material which once shaped by heat and pressure is incapable of being re-processed by further applications of heat and pressure. This is because during the shaping process, cross-links, or crosslinks, are introduced between the molecules which make up the basic resin or plastic. It is these cross-links which make the cured material insoluble and infusible as they stop the molecules slipping past each other and so prevent re-softening. This does not mean to say that these plastics are chemically inert: as will be seen, they are swollen by some chemicals and attacked by others.

Materials such as polyethylene may be cross-linked, after they have been shaped as a thermoplastic, but such materials will not be considered as thermosetting materials in these guides. Those materials which will be treated as thermosets are those which are traditionally known as thermosets - such as phenolics and aminoplastics: these materials are processed by the application of fairly high pressures in processes such as compression molding, transfer molding and injection molding. Also included in this category of materials are epoxides and, molding compositions based on an unsaturated polyester, for instance DMC.

ELASTOMERS

An elastomer (a rubber) may be defined as a material which may be stretched by 100% and on release of the stretching force it retracts, or springs back, rapidly. This elasticity was always associated with vulcanized (cross-linked) rubber and, is due to the structure of that type of material. Now, thermoplastics materials can be made rubber-like.

The long chain molecules on which rubbers are based, prefer to exist in a random, or coiled, configuration. However, the same, long chain molecules are not rigid structures but can untwist (or unwind) around the carbon to carbon bonds, on the application of a stretching force - provided the material is above its glass transition temperature. The chain is not stretched it is simply made to adopt a different configuration, or layout, but it always remembers that it would prefer to exist in a random, or coiled, state. So, when the stretching force is removed, back it goes to a random, coiled state. (A rubber molecule may be likened to a steel spring which, because of the way that it is constructed, may be easily stretched).

The crosslinks serve to stop the chains slipping past each other when they are stretched. If they were not there, then the chains would find it easier to untangle, orientate in the direction of force application, and then move past each other: this would cause the material to separate into pieces.

MATERIALS MODIFICATION

It is generally understood in the plastics molding industry that the materials used, as injection molding materials, may contain small amounts of additives. Such additives may include stabilizers and lubricants. Materials modification usually means the addition of larger amounts of additives. Such additives may include elastomers, flame retardants and fillers.

Materials modification, such as with fibers or with fillers, is extensively adopted with thermosetting plastics, and now with engineering plastics, so as to obtain a desirable combination of properties. It is not done simply to save money as often a molding, made from a filled compound, is the same price as one made from the unfilled plastics material. This is because of the high density of most fillers and because of the .high compounding costs. Many of the fillers used are fibrous fillers as the use of such materials improves properties such as modulus.

REVIEW OF MOLDING METHODS

COMPRESSION MOLDING

Description of process

This is performed on a vertical press, where one half of a heated steel mold is attached to a stationary platen and, the other half to a moving platen. A specific amount of material is loaded into the stationary half and, upon closure of the mold, heat and pressure are applied which shapes the material to the contours of the mold: the excess is forced out between the flash lands in the form of flash. After sufficient time has elapsed for a chemical reaction, or cross-linking, to take place (cure time) the mold is opened and, the component/s removed by a suitable ejection method.

Advantages

Can produce large articles to close tolerances.
Can obtain maximum electrical properties from a material.
When fiber filled materials are used, minimal break-down occurs to fiber lengths therefore maximum strength can be achieved from the molded component.
Capital equipment costs are less when compared to the transfer and injection processes.

Disadvantages

Long cure times for thick sections as cure time is proportional to wall thickness.
Labor intensive process, because of flash removal problems.
Maintenance of concentricity and height tolerances can be difficult as the mold is loaded when open and, and the component shape is contained in both mold halves.
High pressure on cold material causes wear and damage to cores and inserts.

TRANSFER MOLDING

Description of process

This process has also been called one shot, injection molding and like injection molding, the mold is closed and clamped before the material is made to flow and to fill the cavities. The mold has a centrally positioned, loading chamber and the appropriate number of cavities. Each mold cavity is connected to the centrally positioned, loading chamber by a feed system - runner(s) and gate(s). The molding material is loaded into the chamber, pressure is applied to the material by means of an auxiliary or integral ram. Material flow is thus obtained: this allows the mold cavities to be filled with an homogeneous, or plasticized, melt. After sufficient time has elapsed for curing to take place, the mold is opened and the components removed (together with the feed system which may comprise cull, runner and gate). The cull is the cured material in the centrally positioned loading chamber.

Advantages

Can produce complex components, possessing metallic inserts and/or, thin sections relatively easily.
Fast cycle times for thick-walled components possible.
The loading of material into the mold is easier and faster for multi-cavity work.
The amount of flash produced is relatively small.
Gives good control over the thickness of the component.
Can use stiff flow materials.
There is a reduction of finishing costs as there is less flash.

Disadvantages

Increased material usage/wastage (material in the pot/cull, runner and gate.
Increase in wear to mold surface, especially in gate areas, due to material abrasion.
The effect of shrinkage variations results in distorted and/or warped components.
The properties of the molded components are dependent upon gate position, gate size and, orientation of filler in filled materials.

INJECTION MOLDING

Description of process

Most commonly involves the use of a screw and heated barrel assembly to heat soften (plasticize) the plastics material: the heat softened material is then injected into a closed mold. When the cycle commences, the mold closes and heat softened material (plasticized) is injected into the cavity of this closed mold via a feed system (runners and gates): it is injected by the action of the screw moving forward i.e. the screw is acting as a ram. The screw is then rotated. Material is drawn into the barrel, by the action of the rotating screw, and becomes plasticized due to a frictional (as a result of screw rotation) and conductive heat imput. The screw moves back as a result of the pressure generated by conveying the material and the amount of screw movement controls the shot size. When the material in the mold is hard (that is, cure is

11

completed) the mold is opened and the components, and the feed system, are extracted from the mold.

Advantages

Gives a significant decrease in cycle time.
Accurate control of shot weight is obtained.
Consistency from one component to another results.
Action of the rotating screw can help to heat the material uniformly.
The control of flash thickness allows for automatic deflashing.
Cavity filling times are extremely fast.
Removal of gasses and volatiles from the melt occurs.

Disadvantages

The presence of flow lines, or marks, on the components provide failure points and are unsightly.
Production must be continuous because of the hot material contained in the barrel.
More skill required from molding personnel.
Extremely consistent materials (e.g. in terms of flow properties) required.
Wastage of materials in the feed system can be high.
High costs can be incurred, for replacement of screw, barrel and sections of mold, due to material abrasion.

INJECTION COMPRESSION MOLDING

Description of process

Plasticized material is injected from a screw and barrel assembly (a plasticizing unit or, injection unit) into a partially opened mold. When the exact amount of material has been delivered, the mold is fully closed in order to shape the material. When cure is completed, the mold is opened and the shot, the components and the feed system, are extracted from the mold.

Advantages

Orientation of both the plastics material, and the filler, is minimized.
Fewer stresses within molding result.
Better (even) shrinkage of materials/moldings occur.
Freedom from flow marks or weld lines in the final moldings.
Less mold wear than injection molding.
Good surface finish on molded components.

Disadvantages

More expensive and complicated molds than used for injection molding.

Normal injection molding machines require modifications.

Injected volume must be very precise and reproducible.

Reproducible, accurate positioning of the moving platen is critical.

Back-flow down screw must be prevented during compression stroke.

COMMENTS ON HIGH PRESSURE MOLDING

Like other branches of the plastics industry, microprocessor controls are now being extensively used in the thermoset molding industry. This is because the use of such systems, as with the injection molding of thermoplastics materials, permits centralized setting with greater shot-to-shot repeatability. When microprocessor controls are coupled to improved hydraulic systems (for example, which have proportional hydraulics and variable speed pumps) then, faster acting molding systems, which can be energy efficient and accurate, are the result. Setting conditions from previous molding runs can be quickly and easily recalled from plug-in memories. Mold preheating and conditioning, prior to installation in the press, also help to eliminate dead time and thus give faster starting.

When compression molding, microprocessor controls can eliminate the need for pre-pressure and venting by, for example, allowing the selection of slow molding speeds at any point in the mold closing sequence. Slow displacement, and controlled heating, of the material in the molding cavity results so that volatiles can escape. Significant cycle time savings can therefore result.

In compression molding, production modules (cells) of, say, four station units (which consist of four presses driven by one central hydraulic system) are becoming popular. The four presses are all set from one microprocessor control center and automatic loading, feeding and unloading are all readily available. Such production units are mainly used for the molding of relatively small components, for example, UF electrical accessories.

RTM AND PMC

In this book, the molding methods referred to are compression molding, transfer molding, injection compression molding and injection molding. These are the traditional, high pressure molding methods. However, the molding industry is broadening with the growing use of processes such as Resin Transfer Molding (RTM) and with the growing use of Polyester Molding Compounds. (PMC).

13

UNSATURATED POLYESTERS AND STYRENE

Unsaturated polyester resins (UP) consist of a resin dissolved in a monomer, for example, styrene: this gives a liquid system which can be set at room temperatures without volatile emission. High strength is imparted by a reinforcement which is most commonly a form of glass fiber. The products are known as GRP - which stands for glass reinforced plastic or, glass reinforced polyester.

When first introduced, the materials were molded by hand lay up (HLU) molding and then, by spray lay up: such molding methods permitted the production of very large components at low cost (at low cost because there is no molding pressure). However, because of styrene emission problems, and worries about breathing in the styrene, the use of such processes is being reviewed in many factories/countries. In Europe, for example, the limit for styrene emissions will probably be 50 parts per million (ppm). Because of such considerations, closed mold techniques are being actively considered. One such technique is resin transfer molding or RTM.

RESIN TRANSFER MOLDING

Resin transfer molding, or resin injection molding, was developed many years ago but has recently undergone a resurgence of interest. This is, for example, because the use of a closed mold gives a significant reduction in the amount of styrene in the workplace: compared to spray lay up, RTM can reduce the styrene concentration by approximately 75%. The use of two mold halves also means that both sides of a finished molding, have a better finish.

The fibrous reinforcement is placed inside a mold which is then closed and clamped. The catalyzed resin is rapidly injected using a resin injection molding pressure of 400 to 500kPa, 4 to 5 bar or, 56 to 70psi. When the mold is full, the resin gels and then cures (sets) before it is ejected or demolded. Because of the low pressure involved, this process is believed by many to be a cost effective alternative to hand lay up and spray lay up.
Comparison of processes.

Name	Molding complexity	Mold cost	Part cost	Prod vol	Fiber usage range	Typical size limit (t)
HLU	2	1	9	1-200	9	0.5
Spray	1	2	8	1-500	7	1
Cold press	3	6	5	100- 10,000	5	0.2
SMC	4	6	5	100- 20,000	2	0.2
Rotational	7	7	2	500- 3,000	1	0.1
RTM	6	4	4	100- 15,000	8	1
Injection	9	9	1	5- 100,000	1	0.02

In the above table, 9 is the highest cost, complexity or the range of fibers used; 1 is the lowest cost, complexity or the range of fibers used.

POLYESTER MOLDING COMPOUNDS

Polyester Molding Compounds (PMC) are based on unsaturated polyester (UP) resins and reinforcements: other additives include catalysts, fillers, release agents, thermoplastics materials and coloring systems. There are many different types which are nearly always referred to by acronyms.

PMC Acronyms

Polyester Molding Compounds or PMC, is the generic, or family, name.
BMC means Bulk Molding Compound.
CIC means Continuously Impregnated Compound.
DMC means Dough Molding Compound.
GMC means Granular Molding Compound.
SMC means Sheet Molding Compound.
TMC means Thick Molding Compound.
XMC means a continuous roving Molding Compound.
ZMC is a low viscosity compound specifically developed for the injection molding of exterior parts for the automotive industry.

Dry granules/pellets (GMC)

These materials are pelleted materials, just like thermoplastics, and are used in the injection molding industry to produce components for the domestic appliance and electrical accessory market. As the fiber length and concentration is comparatively low, they often contain 10 to 15% of chopped or milled glass fibers, the mechanical properties are not outstanding.

Dough or billet-type compounds

Included in this category are DMC, BMC, CIC and ZMC. Such compounds are used for both injection and compression molding. They often contain 12 to 18% of chopped or milled glass fibers, the size of which is approximately 12.7mm/0.5in. They have better mechanical properties than GMC and find extensive use in the electrical, construction and automotive markets.

Sheet-type compounds

Included in this category are SMC, TMC and XMC. Such compounds contain 20 to 60% of glass fiber and as the glass content is high, and the glass length is long, they have excellent mechanical properties. The glass length is approximately 25.4mm/1in: in the case of XMC it is continuous. Sheet-type compounds have better mechanical properties than dough or billet-type compounds: they find extensive use in the electrical, construction, chemical and automotive markets.

Comments on PMC

PMCs are now widely used, particularly in the transport industry, as compared to thermoplastics, such materials offer better mechanical properties, higher temperature resistance, improved dimensional stability, lower/zero post-molding shrinkage and low flammability. In thirty years, the market in Europe has grown from 500 tons to approximately 0.25 million tons and yet, this is relatively small usage compared to thermoplastics/engineering thermoplastics. It is thought that the variability of such materials hampers their growth.

MOLDING OF THERMOSETTING PLASTICS

Each material will be discussed under a set of standardized headings, so as to facilitate comparison and, the purpose chapter, is to define or explain the headings used.

Common Names

Most plastics materials are known by impressive-sounding chemical names which are unfortunately neither logical nor systematic. The traditional thermosets use the names of the starting materials or monomer so we have (see Table 1 and, Table 2) phenol-formaldehyde, urea-formaldehyde and melamine-formaldehyde. Dough molding compound, based on unsaturated polyester resin, is named because of the consistency of the molding material: epoxides are so named because of the chemical group which they contain. Other nomenclature systems have been proposed but as yet the old-fashioned, illogical system seems to be winning.

Abbreviations

Because of their complex names most plastics materials are identified by abbreviations. There are standards which specify abbreviations (for instance British Standard 3502, or BS 3502) but in this section both standard and non-standard (but widely used abbreviations) will be given. (See Tables 1 and 3).

Alternative Names

This category is necessary as some materials are known by more than one name. For example, epoxides are also known as epoxies.

Some Suppliers

These are not necessarily the manufacturers as like any other commodity, plastics are bought and sold on a world-wide basis: some companies also specialize in making plastics compounds by, for example, adding fillers and in such a case, the resultant compound may be marketed under their own trade name. See table 3 for a list of some abbreviations and trade names/trade marks.

Trade Names or Trade Marks

Where possible the name associated with a particular Company's polymer should be known – this is because in some molding shops, plastics are often only known by their trade name or trade mark. An alphabetical list of some commonly encountered trade names/trade marks, their abbreviations and suppliers, is given in table 4.

Materials properties

Because each of the plastics listed is in effect a generic name for a family of materials (which may differ in chemical composition, degree of cure, etc.) any properties given, or commented upon, should only be regarded as general guidelines. The problem is made even worse by the fact that materials modification with fillers (or filler combinations), is extensively practised with thermosets to obtain a desirable combination of properties, like ease of molding, control of shrinkage, adequate impact strength, cost, etc. Materials are available in a number of grades such as general purpose (GP), improved impact resistance, improved heat resistance and the best electrical properties.

It should also be remembered that if the molding material is shaped by injection molding, this process will introduce fiber orientation. The properties of components produced by, say, compression molding and injection molding will, therefore, be markedly different.

In general, however, thermosets are selected for many applications because of their non-melting behavior on heating, their high heat, chemical and creep resistance, stiffness, surface hardness, dimensional stability and low flammability. Of particular importance is the ability of this class of materials to withstand a load (or stress) at elevated temperatures.

If the service temperatures of glass-reinforced plastics are compared then many engineering thermoplastics appear to outperform thermosets. This is because, for example, the short term limit quoted is an indication of the peak temperature possible under no-load conditions. But if a load is involved, at a temperature, say, as low as 100°C or 212°C, then an entirely different answer is obtained as significant distortion occurs for many thermoplastics. Steam can seriously aggravate the problem for several thermoplastic systems as it causes hydrolysis.

Thermosetting moldings are generally harder and more brittle than engineering thermoplastics and this means that in order to get sufficient strength, wall thicknesses must be relatively large. Because of this, and because of the relatively high densities of thermosets, moldings tend to feel very solid and heavy. In the electrical industry, advantage is taken of the brittleness of thermosets where "knock-out" sections are molded into boxes; if cable entry is required at a particular point the electrician pushes out the appropriate knock-out panel. This is difficult to arrange with thermoplastics.

18

Forms supplied

All of the systems under discussion can be supplied as free-flowing solids – either powders or granular solids (DMC is supplied as a sticky, putty-like material which requires special handling techniques). Grades are usually available to suit a process. Such molding materials may be supplied in a variety of particle sizes (called "grinds") so as to suit a particular molding process. For example, some compression molding grades may be offered in four grinds:

i.   fine grind – for the best surface finish on flat components;
ii. medium grind – a general purpose powder which is also suitable for pelleting;
iii autogrind – a coarsely ground material with few fine particles for use on automatic equipment;
iv fines free – as iii but having no fine particles. Recommended for automatic molding where infra-red preheating is employed.

Injection molding powders are usually free-flowing granular solids – achieved by using a narrow particle size distribution   and by making the particles as regular in shape as possible.   The curing characteristics should be such that the rate of cure at barrel temperatures is very low but reasonably fast at higher (mold) temperatures. Reversion, or a decrease in properties, should not occur if the material is cured for slightly differing times.

Materials handling, packaging and storage

The way in which these materials are supplied will be described together with the anticipated shelf-life if stored under the recommended storage conditions.

When injection molding, the material is simply placed in the hopper, there are no preheating or predrying operations performed. With compression and transfer molding a pre-molding operation is commonly carried out: the two most common processes are pelleting and pre-heating.

A molding powder is turned into a cylindrical pellet (which may be 100mm diameter and 50mm thick i.e. 4in by 2in) by cold compression with a punch and die. The pellet size is adjusted to suit the component and each cavity is fed with the correct number of pellets (usually one or two). Manual material handling is therefore simplified and high frequency preheating made more uniform. Because less space is required to hold the molding charge, molds are smaller and the use of pellets results in less air entrapment (porosity) in the moldings.

Preheating reduces cycle times, mold wear and porosity: it also gives better surface finish and higher mechanical properties. High frequency (HF), also known as radio frequency (RF), preheating is widely used.

Color possibilities

Generally speaking, it is not practical to get clear moldings from thermosets because most molding powders are extensively filled. If the base resin is colorless, both translucent and opaque colors are encountered: what is obtained also depends on the other additives used. Molding powders are supplied precompounded, coloring at the molding machine is not usually performed. The surface appearance of a molding may be changed by in mold coating, by foiling or, even by painting.

Molding conditions

Thermosetting molding powders are shaped by either compression, transfer or injection molding (a brief guide to these processes appears later). These are high pressure molding processes and so the equipment used must be built to withstand these pressures. Even compression molding which uses the lowest pressures, may require 4 tons per square inch (4 tsi or $62MNm^{-2}$) of molding pressure (the area specified refers to the projected area of the molding) although 2 to 4 tsi is more usual. Powders of easier flow may be specified if the pressure available is limited and/or, preheated material may be used.

The mold temperatures are usually below 160°C or 320°F although higher temperatures may be encountered in injection molding feed systems. A range of molding temperatures is usually quoted for each molding material; when compression molding for example, the highest temperatures are used in the automatic molding of preheated material. This is because of the strict control (time and temperature) exercised when molding automatically. As component thickness increases, then the curing temperature must be decreased but the cure time must be lengthened so as to ensure that the component is adequately cured or set. This usually means that the component can be ejected without distortion and that it passes a degree of cure test. With some materials (PF, UF and MF) the gases (volatiles) which are generated on cure will cause serious distortion unless the material is adequately cross-linked.

Mold considerations

The gases or volatiles emitted from these materials should be treated as harmful and adequate precautions (e.g. ventilation) must be taken. Details of mold design and construction, specific to the material under discussion, will be given under this heading together with advice on gate design, component removal and so on. Because many thermoset moldings contain inserts, which need to be loaded into hot molds, automated component-handling devices should be a great attraction. By eliminating manual loading and unloading, the consistency of the moldings should be improved as cycle consistency is introduced and, removing the operator from the area of the hot mold should result in a safer operation for the operator and for the machine. This is because, for example, tool damage can easily occur unless all the flash and the components are ejected at the end of every shot. The action of closing the mold on this very-hard, set material can cause tool hobbing. Removal and inspection are the responsibility of the operators in manual systems: these operations are done under very unpleasant conditions and the chances of mistakes are high with human operators.

Ease of flow

A measure of viscosity, or ease of flow, is given by the cup flow test (specified in standards such as ASTM and BS). Stiff flow powders have cup flow times of greater than 20 seconds, medium flow 13 to 19 seconds and easy flow (soft or free flow) under 12 seconds. A measure of the rate of cure of a particular material, can be obtained by determining the time to produce a blister-free cup. Both ease of flow and rate of cure are affected by changes in temperature. Generally, raising the temperature speeds up the rate of cure but also makes the material easier flowing. Because of this, the total amount of flow possible usually goes through a maximum as the temperature is raised over a material's processing range. The viscosity is shear-rate dependent, but under conditions of constant temperature, it does not vary much with time until the material hardens. Injection grades are easier flowing than compression grades (the two grades have different flow and rate of cure characteristics and are not interchangeable). When judged by a spiral flow test (Mesa, transfer spiral flow) a compression molding grade would give a spiral flow length of 102 to 152mm or 4 to 6in, a transfer molding grade 102 to 254mm or 4 to 10in, and an injection molding grade 254 to 610mm or 10 to 24in.

Shrinkage

Most thermosets, like thermoplastics, shrink on cooling. (Some thermoset materials such as DMC can be made to exhibit very low or even no shrinkage). The figure quoted is usually given in mm/mm or in/in. A range is usually quoted as the actual shrinkage obtained depends not only on the grade of material employed but also on the conditions under which it was molded and the actual molding process involved. For example, phenolics are sold in injection or compression grades and, because injection involves longer flow lengths. these materials must exhibit low viscosity at melt temperatures. However, easy flow materials shrink more (up to 30%) and their use can increase distortion. (Raising the barrel temperature and "breathing" the mold decreases shrinkage).

Resistant to

Where it is stated that a material is resistant to a particular chemical this should only be taken as a rough generalization as chemical resistance will depend upon, for example, the degree of cure and the type of filler employed.

Not resistant to

Because thermosets are classified as being insoluble, infusible materials there is a tendency to think of them as being chemically inert. It is true that they cannot be dissolved without chemical change but some chemicals will attack them and thus cause degradation: they will swell in selected solvents. As half of the molding composition is filler, this has a tremendous effect on properties. For example, organic fillers such as woodflour absorb water and/or are subject to microbiological attack.

Tests for cure or stability

The degree of cure has a great effect on properties but it does not affect all properties in the same way. Altering the degree of cure (by, say, altering the cure time) will not change all properties in proportion. Ideally, the cure conditions needed to optimize a particular property, or set of properties, should be determined by experiment. It is usually assessed by boiling or staining tests, such as chloroform extraction for DAP.

To avoid in-service cracking, caused through high internal stresses, the stability of injection moldings is sometimes assessed by stoving the moldings at elevated temperatures for long times. The moldings in the case of aminoplastics, are judged to be satisfactory if they can withstand 48 hours without developing cracks.

Detection or identification

This refers to identification of the cured molding and, only those tests which can be performed relatively simply and cheaply will be described. When the test involves burning some of the material it should be performed in a fume cupboard and only a small sample should be used. After heating in a flame, take great care if you decide to smell the charred material as the fumes evolved on burning are harmful.

Component finishing or decoration

In many cases the only finishing operation performed on thermoset moldings is one of deflashing. For large components this is often done by hand (using files, linishers,m etc.): smaller components are nearly always deflashed automatically, by rumbling or barreling. This removes the bulk of the flash and the remainder is removed either by hand or automatically. A popular technique relies on conveying the molding through a fast-moving, air stream which contains pellets (for instance of nylon) and these knock the flash away. The need for flash removal, and for flash scar polishing, is a very serious problem as it is expensive, time consuming and, can result in component damage: the inability to use the flash for any useful purpose is also frustrating. As it is nearly always cheaper to dump it than to try reclamation, it makes sense to keep scrap generation to an absolute minimum. Flash removal is more of a problem in compression molding as when injection molding, the flash produced is very thin and can be knocked off by hand: a quick "lick" with a file is often all that is required.

The dust, or trimmings, removed from these materials during finishing operations, should be treated as harmful and adequate precautions (e.g. ventilation, prevention of dust ingestion, skin protection etc.) must be taken.

The brittleness of many thermosetting systems means that jointing by snap-fits is not possible and, as the moldings are cured, jointing by welding is also not possible. The cost of joints in thermosets is very significant. Moldings may be joined together by adhesives, thread cutting screws, rivets, etc. If, for example, rivets are used then problems may arise in production through the inadvertent use of the wrong size of rivet. Special tapping screws have been developed for thermosets. It is now also possible to electroplate some thermosets using techniques similar to those employed for thermoplastics. The marking of thermosets by laser, for example, carbon dioxide lasers, is now also possible. At present, dark lettering on a light background (burning in) and, relatively pale lettering on a dark background (engraving) are both performed.

Other comments

In this section any relevant comments, not covered in other sections, on the molding of a particular plastics material will be included. When considering the molding details of an individual material it is all to easy to forget the object of that particular molding process. The object of molding is to produce moldings of the required quality, in the right quantity, at the right price and, at the right time. This can only be done by careful consideration of each part of the production process. This careful consideration is becoming even more crucial with the growing use of fully automated, molding as robots, pick and place devices, conveyors, assembly machines etc., depend on a constant flow of molded components to maintain production.

The importance of careful and accurate record keeping cannot be over-emphasized.

Typical components

Compared to thermoplastics, the tonnage of thermosetting materials used is relatively small. Approximately 10% of resin use is accounted for by thermosetting resins: these are mainly aminos, unsaturated polyesters, phenolics and epoxides. The non-melting characteristics of thermosets molding materials, the low cold flow (creep), the high shear modulus, the high Youngs modulus (at temperatures up to 300° C, that is 572° F, gives these materials a guaranteed future.

Such materials are traditionally used in electrical applications such as plugs and sockets. In the electrical industry, thermosets can withstand more easily than can thermoplastics, quality assessment tests such as the 850° C, or 1562° F, glow wire test. the highest demand for thermosets is in the electrical engineering and electronic industries.

Household appliances, such as oven-proof kitchen fittings and handles, are steady markets for thermosets, as these materials can withstand high temperatures without distortion while under load and also retain their good surface finish. A traditional area of use for thermosets is in the automotive industry where the performance at elevated temperatures, resistance to hydrocarbon fuels and oils, high compressive strength, stiffness and the high modulus of elasticity are turned to good advantage.

Increases in use, are expected in the domestic appliance area, in tableware, in machinery and vehicle production and, in applications which involve sliding elements, seals and heavy duty pumps. It is the melamine, melamine phenolic and, unsaturated polyester materials (MF, MPF and UP) which are expected to expand their markets most rapidly.

The range of use of thermosetting materials, will broaden as the practice of using blends, or alloys, becomes more common. Such multi-component materials are obtained by using mixtures of resins (binders), mixtures of thermosetting materials and thermoplastics materials, mixtures of thermosetting materials and elastomers, mixtures of thermosetting materials and thermoplastic elastomers and, molding materials obtained by using each of the previous categories with a mixture of fillers or reinforcements.

For example, modification of phenol formaldehyde with an epoxide (PF with EP see table 1) gives moldings which have improved electrical properties at elevated temperatures - such moldings are used in commutator production. The use of an EP

24

resin also improves fiber adhesion, for example, to glass fiber (GF) so, better mechanical properties result. With aminoplastics the blending of urea formaldehyde and melamine formaldehyde (UF and MF), gives moldings which are relatively cheap, light in color and lightfast. By combining a MF with an unsaturated polyester (UP) it is possible to produce materials which on curing give two interpenetrating networks - these are produced as the UP sets by a free radical polymerization mechanism and, the MF sets by condensation polymerization. The moldings exhibit lower shrinkage and post-molding shrinkage than MF alone and, the hardness, compared to UP alone, is increased: fast curing is also obtained. By elastomer (rubber) modification of, for example, PF reinforced with GF, it is possible to produce moldings with high stiffness and which exhibit good damping characteristics. Such moldings are used in the automotive industry as a substitute for die cast metal components.

Unit price

The price given will be based on the assumption that the material is purchased in 1 ton/tonne lots; a tonne is 1,000 kg. If purchased in this way then the plastics material will probably be supplied in 25 kg/55.1lb sacks, with 40 sacks shrink wrapped onto a pallet. When supplied in 50lb/22.6kg sacks, then the material is often supplied in 20 sack (1,000lb) loads. Supplying the material in sacks, or bags, is not the only way that the material can be supplied: considerable discounts may be obtained if the plastics material is purchased in bulk and, supplied in, for example, bulk containers. With some materials, this form of supply may not be acceptable as, for example, excessive water absorption may occur.

The price quoted will be relative to a widely used material, for example, general purpose phenol formaldehyde (GP PF) and this ratio can be found in Table 5; the price of the general purpose material = 1.00.

As plastics materials are bought by weight and sold by volume (in the form of moldings) then, any material with a low density, or specific gravity (SG), has a tremendous advantage. It is misleading, however, to compare the price of one material with that of another, as one should be more interested in seeing how much performance, per unit of cost, is given. However, the cost per unit volume is often more important than the cost per unit weight. To convert to the cost per cubic inch (cu in), multiply the cost per pound (lb) in cents by the SG and then, by 0.0361.

ALLYLS

1 Common name(s)

Allyls; allyl molding materials.

2 Abbreviations

DAP; DAIP.

3 Alternative names

Allylics; allyl molding compounds.

4 Some suppliers	5 Trade names or trade marks
Occidental	Durez
Parr	US Prolam Inc.
Plaskon DAP	Plaskon Molding Materials Div'n.
Polychem	Budd Co.
Poly DAP	US Prolam Inc.
Synres Almoco	DAP = 5000 series
Synres Almoco	DAIP = 6000 series.

6 Material properties

If the monomer diallyl phthalate (or diallyl isophthalate) is heated than it will react or polymerize: the solid reaction product (an allyl resin) can be blended with fillers, peroxide catalysts, etc., to give molding powders which are usually referred to as a DAP (or DAIP) molding material or compound. Because the materials are esters, i.e. products of an acid and alcohol, they are sometimes referred to as "alkyds" (see GPMC) but this practice is not recommended. Use of the initials DAP and DAIP is reserved for the molding compounds which have the same names as their monomers, i.e. diallyl phthalate and diallyl isophthalate. DAP is sometimes referred to as "ortho" and DAIP materials are sometimes referred to as "meta". The isophthalate-based materials are recommended for higher temperature applications.

Both types of resin may be blended with a variety of fillers so as to give a range of grades, some of which have very high impact strengths: achieved by using high concentrations of long glass fiber (GF). Easier processing grades are obtained if the GF concentration and/or length is reduced: can replace some or all of the GF with mineral fillers. Compounds with lower densities, or specific gravities, result if the GF is replaced by an organic-based fiber, e.g. nylon or terylene.

DAP/DAIP compounds are suitable for continuous use in the temperature range 160 to 180°C or, 320 to 356°F. They exhibit excellent electrical insulating properties which are retained under severe environmental conditions (high temperature and high humidity) and, the dimensional stability is exceptional – partly due to negligible after-shrinkage.

DAP (allyls) and alkyd molding materials have, broadly speaking, similar properties; end-users and processors would employ similar processing conditions for both materials. They are both supplied as dry granular solids but DAP materials are more expensive; they do however, retain their excellent electrical properties even under severely humid conditions and also resist biological attack.

7 Forms supplied

Supplied in granular form suitable for automatic processing. Where high strength is required, long glass fibers are incorporated and such compounds may be supplied in flake and nodular form.

8 Materials handling, packaging and storage

The material is of a granular nature and is relatively dust free. It is often supplied in plastic drums, which contain 25 to 30kg/55.1 to 66.1lbs, and if stored under normal cool conditions (10 to 15°C or, 50 to 59°F), a shelf life of 12 months or more can be expected. Allow the material to reach workshop temperature before opening the container e.g. keep in molding shop unopened for 12 hours before use.

Flake type materials are supplied in cartons and handling procedures must be carefully considered due to the nature and sharpness of the flakes. Flake materials must be handled using protective clothing, e.g. gloves.

DAP's have a hard, tack-free nature and therefore can be automatically transported without the risk of agglomerating into lumps.

9 Color possibilities

The natural color of base compound is water white and with the common fillers used, a light, beige colored material is obtained. This enables a wide range of colors to be produced, e.g. reds, blues, whites, etc.). Color matching is possible but is dependant upon quantity required. Color stability under U.V. light and heat is affected by the type of pigment used.

27

Can be compression, transfer or injection molded. However, for injection molding, grades specially developed for injection molding are used, for example, the "NB" (New Briton) type grades. When compression molding, high frequency preheating is regularly used: preplasticizing units are only incorporated in the production unit, if the production quantity can justify the cost of the unit.

The powder and nodular materials are readily pelleted using pelleting pressures of 1 to 2 t.s.i. (15 to $31MN/m^2$). Flake type materials are not usually pelleted and various hand charging systems are used to obtain correct shot weight.

Typically preheating temperatures of 70 to 80°C (158 to 176°F) are used. These temperature are used so as to control the material's viscosity and thus, overcome problems such as driving out unwanted air from the mold cavity. However, quite frequently cold pellets can be used with satisfactory results.

If preplasticizing units are used preheat temperatures of 70°C or 158°F are recommended. The press cycle is arranged so that the preplasticized slug is not allowed to stand, or cool down, prior to being positioned in the mold cavity.

Automatic compression molding is possible but is dependent upon production volume. If DAP/DAIP needs to be molded after epoxides or phenolics, it is advisable to use a UF or MF first, so as to remove any contamination and also to clean the surface of the mold.

The transfer molding principles used for phenolics can be readily adapted to produce components from DAP and DAIP. As a result of DAP's easier flow characteristics, smaller sized runners and gates tend to be used than those required for phenolics. When injection molding DAP/DAIP, typical screw designs having an L/D ratio of 10 to 15:1, and a compression ratio of 1:1, are used: a back flow valve is not required. As a rule, the screw is only designed to transport the material to the front end of the cylinder.

Plasticization of the material takes place within the front/nozzle section of the barrel (i.e. last 10%) and, melt preparation is achieved by the combination of barrel temperature and screw back pressure. Although the mold surface abrasion is less than that encountered with PF's, the amount of wear is dependant upon the type of fillers used. The type of components molded tend to be complex (e.g. connectors and coil formers) and as flatness and close tolerances are required cooling jigs are used to achieve the required quality.

11 Mold considerations

The most popular design used for compression molding is the semi-positive, vertical flash type. Depending upon the nature of material used, the clearance between punch and die varies. For example, for flake the gap is 0.15 to 0.18mm or, 0.006 to 0.007in. For granular materials, clearances of 0.05 to 0.13mm, or 0.003 to 0.005in, are used. Vacuum venting is not essential for injection molds as air escapement can be accommodated by normal venting techniques. For example, the width of vents is 2 to 6mm (0.079 to 0.236in) and, the depth is 0.03 to 0.05mm (0.001 to 0.002in). Typical runner diameters, used for injection molds, are 4 to 6mm or 0.157 to 0.236in. The types of gates used for DAP's are edge, tab, film and submarine. The steel used for compression, transfer and injection molds is of the nickel-chrome alloy type, e.g. H.13 or D2, which is hardened to 62 Rockwell C. Most molds are chrome or nickel plated so as to protect the mold surface and improve mold release. The demolding strength of DAP/DAIP is good and, standard ejection methods are employed to obtain distortion-free components.

12 Ease of flow

This can be a relatively easy flow material. Due to the high cost of this material, and the material's easy flow characteristics, runner sizes under the quoted diameters, for injection molding, can be used provided the shape and complexity of molded article has been taken into consideration.

13 Shrinkage

The shrinkage for both materials varies according to the flow direction. For example, along the flow, the shrinkage ranges from 0.1 to 0.2% for both DAP and, DAIP. Across the flow, a more usual figure is 0.6 to 0.8% for DAP and, 0.35 to 0.6% for DAIP.

After shrinkage for both materials is approximately 0.05%; flame retardant grades exhibit similar shrinkage characteristics.

14 Resistant to

These materials are resistant to a wide range of chemicals, e.g. 30% sulfuric acid, 10% nitric acid, 25% ammonium hydroxide, gasoline/petrol, paraffin, lubricating oil, diesel, hydraulic brake fluid, acetone and, chlorinated hydrocarbons (such as trichloroethane and carbon tetrachloride; also resistant to conditions of high humidity. Alkyds can be hydrolyzed or saponified, DAP's cannot. Can withstand continuous exposure to 160° C or 320° F and DAIP can withstand 180° C or 356° F. These figures are not quite as good as epoxy, or epoxide, material which may be formulated to withstand 220° C or 428° F.

15 Not resistant to

Sodium hydroxide: some grades have poor resistance to 10% solutions. 10% oxalic acid can attack some grades.

16 Tests for cure or stability

Reflux part, or all, of a molding in boiling chloroform for 3 hours, carefully remove from the sealed apparatus, allow the molding to breathe or stand and then, examine the surface for evidence of serious attack, e.g. cracking, swelling and cracking. Check Barcol Hardness (nos. 60 to 70) before and after chloroform boil – allow 12 hours to elapse, after removal from chloroform, before testing. A dense, fully cured part will show very little or no effect from solvent treatment (slight roughness on a normal surface). A slightly undercured part will show whitening and/or swelling, whereas an undercured part will show cracks and holes or, be disintegrated by the solvent.

17 Detection or identification

These are usually high density materials which have densities in the range 1.73 to 2.08gcm^{-3} or, 1.0 to 1.2oz/cu.in. (the actual density will largely depend on the filler type and concentration). Such dense moldings sink rapidly in water and, in saturated magnesium chloride. Both DAP, and DAIP, are often used to produce very complex components which have a dense solid feel. Typically the components are molded to very high accuracy as they are used in very demanding applications: may be used as insert/outsert moldings. When placed in a flame the component burns with a bright, sparkling white flame and, the flame itself is tinged with blue. Burns gently in the flame and when removed from the flame, white smoke is produced with a sweet odor. Edges of burnt component show black (with grey surrounds). Both DAP and DAIP behave similarly when placed in flame.

Materials possess a hard surface and are difficult to scratch with a sharp object. When dropped, the moldings show good impact resistance. DAP may be seen on very delicate components having wall sections as low as 0.25mm or 0.01in.

DAPs can sometimes be identified by their surface finish as they are not renowned for their high gloss finishes.

18 Component finishing or decoration

For optimum properties, post-bake the moldings in an air circulating oven at 150 to 190°C, or 302 to 374°F, for periods of 2 to 16 hours. DAP moldings can be machined and drilled by conventional methods. DAP moldings are not as scratch resistant as PF's, or epoxides, so after tumble deflashing (with walnut shell flour) the surface can become marred. Critical components are therefore usually deflashed by hand. Due to the low production volumes of many moldings, hand deflashing tends to be preferred.

19 Other comments

Although these materials are similar to other thermosets, it is the consistent performance under widely varying environmental conditions that make these materials extremely attractive.

Inert to fungus growth and, no "brown powder effect" on silver relay contacts.

20 Typical components

The diallyl phthalate materials (DAP) are better than PFs under conditions of dry and wet heat. Diallyl isophthalate materials (DAIP) are more expensive than DAP but, have better heat resistance: they can withstand continuous operating temperatures of 180°C or 356°F (DAP is approximately 160°C or 320°F). Allyl materials are superior to PFs in their tracking resistance and, in the color range available. They show a lower shrinkage on cure than PFs but, cracking around inserts can occur due to the lower elasticity of this type of material. DAP-type materials, are used in very demanding electrical/electronic applications where, retention of electrical and mechanical properties under extreme conditions, is of paramount importance, e.g. connectors, potentiometer housings, switches, relays, circuit breakers and coil bobbins. Widely used in military applications where absolute reliability is demanded.

21 Unit Price

The unit price relative to PF (General Purpose Grade) is as follows:

DAP General Purpose grade   8.80
DAP Flame Retardant grade   8.80
DAIP General Purpose grade   13.60
DAIP Flame Retardant grade   13.60

The price relative to GPPF can also be found in Table 6.

SUGGESTED PROCESSING CONDITIONS FOR ALLYLS.

Process	Setting	Unit	Value	
			DAP	**DAIP**
Compression	Preheat temperature	°C	50 to 70	80 to 100
	Tool temperature	°C	150 to 170	150 to 170
	Pressure	MNm^{-2}	5 to 20	5 to 20
	Curing time	s	30 to 60	10 to 30
Transfer	Tool temperature	°C	150 to 180	150 to 180
	Transfer pressure	MNm^{-2}	10 to 70	10 to 70
	Curing time	s	90 to 150	90 to 150
Injection	Cylinder temperature	°C	60 to 70	60 to 70
	Nozzle temperature	°C	80 to 90	80 to 90
	Melt temperature	°C	105 to 115	105 to 115
	Screw temperature	°C	5 to 85	65 to 85
	Tool temperature	°C	150 to 180	150 to 180
	Injection pressure	MNm^{-2}	50 to 120	50 to 120
	Back pressure	MNm^{-2}	2 to 12	2 to 12
	Curing time	s	15 to 60	20 to 60
	Injection time	s	3 to 15	3 to 15

Process	Setting	Unit	Value	
			DAP	**DAIP**
Compression	Preheat temperature	°F	122 to 158	176 to 212
	Tool temperature	°F	302 to 338	302 to 338
	Pressure	psi	225 to 2,900	225 to 2900
	Curing time	s	30 to 60	10 to 30
Transfer	Tool temperature	°F	302 to 356	302 to 356
	Transfer pressure	psi	1,450 to 10,150 for both	
	Curing time	s	90 to 150	90 to 150
Injection	Cylinder temperature	°F	140 to 158	140 to 158
	Nozzle temperature	°F	176 to 194	176 to 194
	Melt temperature	°F	221 to 239	221 to 239
	Screw temperature	°F	149 to 185	149 to 185
	Tool temperature	°F	302 to 356	302 to 356
	Injection pressure	psi	7,250 to 17,400 for both	
	Curing time	s	15 to 60	20 to 60
	Injection time	s	3 to 15	3 to 15.

DOUGH MOLDING COMPOUND

1 Common name(s)

Dough molding compound

2 Abbreviation

DMC

3 Alternative names

Premix; bulk molding compound (BMC), polyester molding compound (PMC)

4 Some suppliers	5 Trade names or trade marks
BASF	Palatal (resins only)
Bayer	Leguval (resins only)
BIP	Beetle DMC
ERF	ERF DMC
Fiberglass	Dion and Vibrin (resins only)
Freeman Chemicals	Freemix
Mia Chemicals	Miapol (resins only)
Orkem	Norsomix
Scott Bader	Crystic (resins only)
SIR	Sirester (resins only)

PMC compounds

Ashland Chemical	Aropol and Hectron
Chromos – Ro – Polimeri	Esterform
Dr. Illing	Illandur
Haysite Industrial Plastics	Haysite
Isola Werke	Durapol
Orkem	Norsomix
Pheonix	Keripol
Resart	Resatherm
Toyobo	Vyloglas
Ubbink Nederland	Uromix

6 Material properties

The term "ester" is given to the reaction product of acids and alcohols, so a polyester is a reaction product (a resin) which coins many ester groups. Such resins are often obtained by reacting together propylene glycol (1, 2-propane diol), maleic anhydride and, phthalic anhydride. This toffee-like material, is then dissolved in styrene: this acts as a solvent but it is capable of reacting with the resin when catalyzed or initiated. The resin solution may therefore, be cast and set inside a mold as it is a comparatively thin liquid. However, to get high strength products, fibrous fillers such as glass must be added; to reduce shrinkage and cost, particular fillers, such as calcite or dolomite, are also added.

The resultant mix now has a putty-like, or dough-like, consistency and hence, may be known as a dough molding compound.

To get economic production of high strength, glossy products from these putty-like, polyester molding compounds (PMC) compression, transfer or injection molding must be employed. In its simplest form, DMC consists of resin, chopped glass (approximately 6mm, or 0.236in, fiber length), mineral filler, mold release agent and catalyst. These are thoroughly mixed together to a dough-like consistency in a mixer; Z blade mixers are used but plowshare types are faster. The glass is added last and care is taken to avoid breaking down the fibers (by either reducing the length or, by causing filamentization). The catalyst, which cause heat setting to occur, is a heat activated, organic peroxide such as t-butyl perbenzoate; zinc stearate is a commonly used lubricant. The resin is normally a styrenated, unsaturated polyester (UP) but the monomer (styrene) may be replaced by others, such as diallyl phthalate. The compounds may be formulated to maximize a particular property, for example, flame resistance, impact strength, etc..

Originally, bulk molding compound (BMC) was different from DMC in that it was formulated to give moldings of improved quality and finish. Isophthalic resins are used to give better hot strength and stability. Resins are chemically thickened, with magnesium oxide, so as to reduce filler segregation and also, to improve surface finish by reducing porosity. Large shrinkage reductions are obtained by adding a thermoplastic powder, like PS or PE, as a mix ingredient. Nowadays the distinction between DMC and BMC has become blurred and often, the terms are used interchangeably although, strictly speaking, they are different. Thick molding compound (TMC) results if the glass fibers are not degraded during compound manufacture. Such lack of fiber degradation gives products with superior strength properties, comparable to sheet molding compound (SMC), so its manufacture and properties are very different from DMC/BMC.

7 Forms supplied

DMC is available as an extruded rope or in "dough" form. Ropes enable molded charges to be measured by length so eliminating weighing: typical rope diameters are 25, 38, 50 and 75mm or, 1, 1$^1/_2$, 2 and 3 inches. The dough form, is supplied as 'cheeses' (weighing about 15kg, or 33lbs, upwards) and each cheese is often wrapped in PE. A batch of cheeses may weigh 120 to 1000kg (265 to 2240lbs) and, is supplied inside a box, or keg, so as to minimize styrene loss. Nylon or cellophane film is often used instead of PE to prevent styrene evaporation.

8 Materials handling; packaging and storage

Most grades of DMC contain styrene monomer so avoid skin contact with this type of material; use polyethylene gloves, on top of cotton gloves, and only use this type of material in a well ventilated area. Do not smoke when these materials are being handled or, if the fumes from them are present in the environment. The dough is tacky and can be easily broken down for charge weighing; if manual handling is used, then a rope form may be more convenient to use. Do not leave the containers open as styrene monomer will be lost and, the sticky material will become contaminated by picking up dirt and dust. The loss of monomer, will make the workplace uncomfortable and will lead to molding problems if, the loss is excessive. Just storing the material in a PE bag will not stop styrene loss. The material should be packed inside PE/PA laminates and then, inside a box or keg (of fiber or steel). If high storage temperatures are avoided, a storage life of three months at 20°C, or 68°F, may be reasonably expected (this depends on storage conditions and formulation – including color). The DMC material will slowly cure on storage (precure) to eventually give a rock-hard product. The smell will either change to that of bitter almonds or, no apparent change in smell will occur. Moldings made from precured material can have markedly inferior properties: increasing the cure time will not help to improve the properties if precured material has been used. Adopt a strict Fi-fo policy so as to minimize problems caused by excessive material storage.

9 Color possibilities

DMC in a range of opaque colours which, on molding, can give a glossy attractive finish. The surface is hard but not as hard as PF. DMC is usually seen as glossy, pastel-colored moldings. The natural color is beige but this depends on the fillers used. It is important to get good color dispersions during compound manufacture as, you cannot color on the machine: the work done, during color dispersion, would cause excessive fiber degradation. Because of the white fillers normally used, black colors are more difficult to obtain. To improve the hardness of the final molding, liquid coatings may be sprayed 'in-mold' and, powder mold coating is now being actively developed. The powder is based on UP resins and, is applied with electrostatic powder spray equipment on to the hot mold here it melts, flows, forms a film and polymerizes. After 20 seconds molding may begin. Such "in mold-coating" (IMC) is more common with SMC; MF resins are also used for IMC materials.

10 Molding conditions

DMC is widely processed by injection molding. In-line screw machines are often used as such machines are then available for processing other materials when the DMC job finishes. Plunger machines are used, but are more restricted in their use: they give less fiber degradation and are useful for large moldings as they are relatively cheap.

On screw machines, doughs are fed into the barrel from a stuffer box; an hydraulically driven piston, applies pressure only while the screw is turning. Such forced charging can lead to fiber degradation as each turn of the screw cuts off one layer of compacted material. Fiber degradation occurs by filamentization of the fiber bundle and, by the breakdown of filaments into elements of lower aspect ratio. Fiber degradation decreases, with an increase in barrel temperature and, with an increase in gate size or diameter. Degradation in the feed system, particularly when passing through the gate, is very serious.

To minimize fiber degradation, injection-compression molding may be used. This process also reduces fiber orientation, thus resulting in a more even fiber distribution and, its use also improves the dimensional accuracy of the resultant moldings. The mold may be designed to cut off the gate aperture – which allows the use of large cross-sectioned gates that aid flow and reduce fiber degradation. Finishing costs are thus reduced and scrap levels are kept down.

To stop material leaking down the screw, during injection, a ring check valve should be fitted. Such a valve should be designed to stop fiber degradation and must be inspected regularly so as to see if it is wearing excessively. If the valve wears excessively (keep the clearance below 0.25mm or 0.01in), then precure problems will occur. The screws used for DMC, do not wear as much as the valves/check rings. Keep the screw rotational speed low, less than 50 rpm, and use a deep flighted screw with a compression ratio of approximately 1:1. As the original DMC compound is soft, barrel temperatures can be comparatively low (20 to 60°C or, 68 to 140°F). Temperatures greater than 60°C, or 140°F, can cause precure; up to a temperature of 60°C, the materials are relatively stable, and may be usually held in the barrel for up to 24 hours but, it is good practice to purge out for overnight stops. Thorough cleaning of the cylinder, or barrel, assembly can usually, only be completely performed by the use of a solvent.

Injection molding gives a better finish than compression molding – because of the time necessary for mold closure in compression. This can be 5 seconds and in that time a skin forms. On pressure application, this gelled material (or membrane) is ruptured and is then dispersed throughout the molding; it then rises to the surface to give a characteristic appearance of porosity. Injection must be completed in less than 2 seconds as the speed of cure is fast. Keep the back pressure as low as possible.

Do not lubricate molds, especially with Carnauba wax, as the use of such lubricants can decrease the hot strength of polyester molding compounds. Lubricant use may, however, prove necessary in the flash-land area.

Preheating is not common with compression or transfer molding. However, if required, try a preheat temperature of 80°C (176°F), but then transfer the hot charge from the preheater quickly, for example, in less than 10 seconds. Do not use smoking/fuming preforms, as the cured material may cause mold damage. Initial mold temperatures should be in the region of 160°C (320°F), but adjust the temperature to suit the job in hand – as high as possible without precure, or without the formation of heavy flash or porosity. Use the stiffest flow material possible and, the highest pressures (consistent with flash cut-off); this will give the densest moldings with the best electrical and mechanical properties. The sticky nature of the material creates contamination and handing problems, especially with multicavity work. Weigh the charge and load the mold with one lump of material; charge positioning is not as critical as for SMC.

Curing times may be comparatively long (compared with injection molding) because of poorer heat transfer. Transfer molding can offer some advantages, such as for insert positioning: the auxiliary-ram type, allows low pot temperatures and therefore, gives better material control.

BMC is less tacky than DMC and has a higher viscosity: it is used as a compression molding material.

11 Mold considerations

Injection molds must be designed to minimize fiber damage and orientation. Fiber orientation is affected by the type and position of gate used – use fairly large gates to minimize fiber orientation and degradation. Runners should also be as large as possible and abrupt changes in direction should be avoided: it is better to sweep the material around bends in the feed system rather than use sharp, and abrupt, changes as is the practice for thermoplastics materials. However, it must be accepted that there will be fiber degradation through the feed system and that the moldings will be anisotropic. If the melt is divided (for instance by a core) then, poor weld strength can be a problem. Warm-runner molding, and extended nozzle molding, reduce waste and can reduce cycle times – this is because sprue curing is often the rate-determining step as sprue sizes are large. When using extended nozzles, a separate control is required to prevent precure within the nozzle due to heat absorption from the mold.

Venting, because of the fast mold filling employed, is extremely important in DMC injection molding, especially in dead-ends and thick sections furthest from the gate. Vacuum extraction must be used on many jobs. Molds should be well polished and hard chrome plated, to Rockwell 60C, so as to withstand wear: ion implantation after chrome plating dramatically improves wear resistance. Design the mold to take account of the low hot strength of polyesters: use the maximum allowable draft (2 degrees) and, use large and numerous ejectors – 25% more than for other thermosets. Gate thicknesses (the depth) should be not less than 0.5mm or 0.02in: back tapers are used for edge and tab-type gates to avoid gate vestige (scarring when gates are removed). For low profile materials, the taper used on sprue brushes should be greater than 3 degrees.

Vertical flash, semi-positive type compression molds are preferred, as the use of such molds enables direct pressure to be applied to the material: the thickness of the molding is controlled by pressure pads mounted at the sides of the cavities.

When using horizontal flash, semi-positive type molds, particular attention should be given to the width of the lands used. (They should be approximately 1.6 to 5mm wide or 0.063 to 0.197in). As DMC cures very quickly, problems can arise when material is trapped and cures on the land surface; this prevents further flow of material across the land and, subsequently restricts the amount of pressure that can be applied to the material in the cavity.

When using fully positive type molds, porosity in moldings (caused by venting difficulties) can be a problem. Due to the material's low viscosity, the closeness of fit between punch and die must be controlled (a typical clearance is 0.08–0.18mm or, 0.003–0.007in).

As the material exhibits very low or nil shrinkage, the correct positioning of ejectors, the use of a good mold surface finish and, the absence of undercuts is essential.

Molds possessing side cores can create weaknesses in moldings owing to separation of flow and should therefore be avoided if, at all, possible. Moving cores, that are incorporated in molds, should be designed with minimal clearance and adequate support so as to prevent the formation of flash; this will cause seizure or sticking during side core movement.

12 Ease of flow

Injection pressures of up to 200 MNm^{-2} (29,000 psi) are usually provided on thermoset injection machines. Because of the easy flow characteristics of DMC (BMC is much thicker), lower pressures of around 35 MNm^{-2} (5,075 psi) may be used. Machines may therefore be fitted with large capacity injection units – which is useful as DMC components are of heavy cross-section and need a large shot weight.

13 Shrinkage

This depends on the formulation used: the shrinkage is essentially a function of the type/quantity of the thermoplastic used, on the filler used and, on the concentration of the glass fiber. A general purpose grade would have a shrinkage of approximately 0.15% (0.0015 in/in or mm/mm) and, the best low profile grades have no shrinkage at all. No post molding shrinkage is observed on the standard 48 hour test at 100°C (212°F), or on other grades commonly available. Low profile grades may show slightly higher water absorptions than GP grades.

14 Resistant to

Outdoor exposure: on prolonged exposure (several years) there may be some loss of color and gloss but, mechanical properties are usually retained. Moldings can withstand high temperatures (up to 160°C or, 320°F) for long periods and 200°C (392°F), for short term exposure. DMC resists water, alcohols, aliphatic hydrocarbons, detergents, greases and oils.

15 Not resistant to

Ketones and chlorinated hydrocarbons; the resistance to aromatic hydrocarbons, acids and alkalis is also not very good and worsens as the temperature is increased.

16 Tests for cure or stability

On removal from the mold examine the molding for obvious defects, such as a lack of rigidity, cracks, porosity, surface blisters and/or swelling and a lack of flatness or warping. Measure the Barcol Hardness, boil in water for one hour, allow to cool thoroughly and then re-check the hardness for under-cure.

17 Detection or identification

DMC is heavily loaded with filler at a filler to resin ratio of around 2:1 and the fiber content can range from 10 to 25%. Densities are therefore high (e.g. $1.7gcm^{-3}$ or 0.982oz/cu.in) and, moldings sink rapidly in both water and in saturated magnesium chloride. Because of the high density, and the thickness of the components, moldings feel very solid and heavy. The surface can be of a very high gloss which is diffi- cult to scratch with a finger nail but can be marred with a knife blade. Large chips can be easily removed with a knife from the molded edge. Moldings are highly colored through- out but now, as inmold coloring (IMC) is possible, the color may only be skin deep.

The ratio of organic:inorganic material may be obtained by heating a weighed sample at 625°C (1157°F) in air. This will burn off any thermoplastic additives and the UP. If required, a calcium carbonate type filler may be dissolved out with dilute hydrochloric acid and the glass fiber, still in fibrous form, will be left. This is a useful technique for studying fiber degradation.

When placed in a flame the molding burns with a yellow flame which when extinguished gives off white smoke and, an organic smell of styrene. When struck very sharply, the mold- ing can fracture relatively easily and the crack may penetrate right through the section.

18 Component finishing

As gates are large, and the material is tough, pneumatic side cutters may be needed for cropping. If machined, be careful not to ingest, breathe in, the dust; use very sharp tools, a well ventilated area and wear dust masks. Drills should be tung- sten tipped. The flash on injection, and on transfer, moldings may be easily broken away but, on compression moldings, it is much thicker and difficult to remove. Hand finishing is widely used but, great care is needed to get a good consistent finish. Remove as much flash as possible when the molding is still hot as this is easier. Some electrical moldings are wiped with an air-drying oil to enhance the appearance and to remove dust caused by filing. Moldings may be painted without abrading the surface but degrease them first: they are sometimes stoved after painting (typically 160°C (320°F), for 30 minutes) and can be coated to provide EMI shielding.

19 Other comments

DMC is widely used for cookware – before use, bake at the highest temperature envisaged, say 180°C (356°F), for 15 minutes so as to remove traces of volatiles (styrene included). As styrene is soluble in hot water, fill with water and heat again before use. This procedure is suggested, because the human palate is very sensitive to traces of styrene. During processing keep the molding area well ventilated so that the styrene concentration is kept low.

20 Typical components

These are highly filled, reinforced materials which have a high flexural modulus and a low coefficient of thermal expansion – similar to metals. DMC is used where precision, strength, attractive appearance and resistance to corrosion and high temperatures, at an economic price, are required. As well as its use in microwave cookware and mid-temperature range oven cooking, DMC is commonly seen in and around the home, as sandwich toaster components, shrouds for domestic irons, etc.. It is also used in office machinery for things such as machine covers because of its good impact resistance and reduced flammability rating – it can have a rating of UL94–VO. DMC has excellent dimensional stability and good water resistance. It is now being used in automotive applications, such as headlight reflectors, and can be made conductive by adding around 5% carbon black although shielding from electromagnetic radiation is best achieved in other ways, such as by using metallized fibers. DMC is also widely used in the electrical industry, for instance for meter boxes and lids. PET fibers are now being used for toughness and resilience: such fibers may hold a molding together when it is struck.

21 Unit Price

The price relative to PF (General Purpose Grade) is as follows:
General purpose grade	1.48
Arc and track resistant grade	1.56
Low profile grade	1.55

The price relative to GPPF can also be found in Table 6.

SUGGESTED PROCESSING CONDITIONS FOR DMC

Process	Setting	Unit	Value
Compression	Preheat temperature	°C	80
	Tool temperature	°C	120 to 160 optimum (140 to 150)
	Pressure	MNm^{-2}	3.5 to 14
	Curing time	s	1mm to 25s/2.5mm to 45s
Transfer	Tool temperature	°C	150 to 170
	Transfer pressure	MNm^{-2}	14 to 28
	Curing time	s	10 to 90
Injection	Cylinder temperature	°C	20 to 60
	Nozzle temperature	°C	20 to 40
	Melt temperature	°C	20 to 50
	Screw temperature	°C	Keep cool
	Tool temperature	°C	140 to 160
	Injection pressure	MNm^{-2}	27.5 to 80.0
	Back pressure	MNm^{-2}	<0.3
	Curing time	s	20 to 90s

Process	Setting	Unit	Value
Compression	Preheat temperature	°F	176
	Tool temperature	°F	248 to 320 optimum 284 to 302
	Pressure	psi	500 to 2,000
	Curing time	s	0.039in to 25s /0.098in to 45s
Transfer	Tool temperature	°F	302 to 338
	Transfer pressure	psi	2,000 to 4,000
	Curing time	s	60 to 90
Injection	Cylinder temperature	°F	68 to 140
	Nozzle temperature	°F	68 to 104
	Melt temperature	°F	68 to 122
	Screw temperature	°F	Keep cool
	Tool temperature	°F	284 to 320
	Injection pressure	psi	4,000 to 11,500
	Back pressure	psi	<50
	Curing time	s	20 to 90s

EPOXIDES

1 Common name(s)

Epoxide molding compounds; epoxies.

2 Abbreviations

EMC; EP.

3 Alternative names

Epoxy molding compound; epoxy; ethoxyline resins

4 Some suppliers

5 Trade names or trade marks

Some suppliers	Trade names or trade marks
Bayer	Lekutherm (EP resin)
Chemroy Canada Inc.	DER (EP resin)
Ciba Geigy	Araldite
Ciba Geigy	Neonite – long glass fibre based, CPC
Emmerson & Cuming Inc	Eccomold
Kingsley & Keith	Eposir (EP resin)
Menzolit	Menzolit Werke
Morton Chemical Div.	Polyset
Plaskon Molding Materials Div.	Plaskon
SPREA	Eponac
Synres Almoco	3000 series

6 Material properties

If epichlorohydrin is reacted with bisphenol A (also known as diphenylolpropane) in the presence of an alkali, then syrups (or brittle resins) of low molecular weight are produced. Such, low molecular weight intermediates have a backbone structure of a polyether with hydroxy groups (–OH) along its length and, a reactive glycidyl ether group (epoxy group) at each end: that is, they are complex diglycidyl ethers. To produce molding powders, such resins are blended with fillers (e.g. glass and mineral fillers), a mold release agent (e.g. zinc stearate) and, a hardening system.

There are two principal classes of hardeners, or curing system,s and these are amines (e.g. diaminodiphenylmethane – DOM), or, acid anhydrides (e.g. phthalic anhydride). Chain extension and cross-linking, which produce the thermoset structure, proceed via addition-type reactions so that volatile by-products are not produced and low shrinkages on curing, or hardening, are obtained.

The polyether structure, of the cured resin, resists many forms of chemical attack and the moldings are therefore relatively inert. This type of structure means that moldings can also withstand relatively high temperatures: rigidity decreases only slightly with rising temperatures, up to 150° C or 303° F, and the moldings exhibit good creep resistance at elevated temperatures. Cured materials have excellent electrical properties even under extreme climatic conditions.

Epoxy novolaks are made in a similar way to bisphenol A resins but are based on novolak resins having a multiplication of phenolic groups instead of bisphenol A type groups. Such resins when cured with an aromatic amine, have the best chemical resistance.

An advantage of EMC is that its properties are unaffected by the length of cure time, provided a minimum is exceeded. Components possessing thick and thin wall sections are uniformly cured throughout as the thin sections do not become overcured while waiting for the thicker sections to cure: this cannot be said for other thermosets.

7 Forms supplied

Available as granular solids which are usually packed in 25kg/55.1lb plastic sacks or drums.

8 Materials handling; packaging and storage

Can have a short storage life (e.g. 3 months), so strict stock control must be exercised. Amine cured materials will have a shorter shelf life, as they react at lower temperatures. Only keep sufficient for about 1 month's use and store under clean, cool conditions – refrigerated if possible. Allow the unopened container to reach workshop temperature before opening, e.g. keep the sack or drum in the molding shop for 12 hours before opening. Avoid dust ingestion from this material and, do not handle the molding compounds with bare hands.

Material is non-hygroscopic, so predrying is not usually necessary. Normally pelletized for compression and transfer molding. Can be autofed for pelleting operation. Long fibre, glass filled grades are pelletized in molds heated to 40 to 60° C, or 104 to 140° F, and if not molded soon after pelleting they should be stored in a refrigerated area.

9 Color possibilities

Traditionally, EMCs were cured by amine hardeners, giving

dark unstable colors, which were often pigmented black. Acidic hardened EMCs can be lighter colored but in practice are often not; therefore only a limited color range is available, for example, offering black, red, brown, blue and green. Special colors available at a price.

10 Molding conditions

Molded by compression, transfer and injection molding. Long glass fiber (GF) filled grades are mainly compression and transfer molded.

Compression molding, as with other thermosets, produces moldings with the highest strength, particularly with the long GF filled materials. The majority of EMCs used for compression molding are pelleted and preheated, or preplasticized, prior to being loaded into the mold.

For long fiber grades it is not advisable to pellet in advance, as swelling of the pellet takes place. This material has particular problems for transfer molding, as pellet size to transfer pot size, is of major importance. Most pellets are high frequency (HF) preheated to 80°C/176°F, but one must ensure that the center of the pellet is at 80°C (176°F) before transfer to the mold occurs.

Preparation of EMC, in preplasticizing units, is often employed – particularly for the long fiber, GF filled grades. In order to prevent damage to the fibrous filler, preplasticizing units having short deep flighted screws of >51mm or 2in, diameter are used. (A typical L/D ratio is approximately 13:1). Use of the preplasticizing unit, gives shorter cure times and higher strength components, compared to production with the HF preheating method. Typical temperatures for preplasticized slugs are 80 to 100°C (176 to 212°F); when produced the slugs should not be allowed to stand, or cool down, prior to being placed into the mold.

Transfer molding machines with adjustable filling rates (i.e. variable transfer ram speeds) are useful (particularly for encapsulation purposes) as a slow, laminar melt flow, prevents 'curing off' during the transfer process. Better control over the melt viscosity is obtained and, an efficient bond between components also results. As with compression molding, the forces used to create material flow are considerably less than with other thermosets and therefore, better utilization of press capacity is obtained, e.g. more cavities/mold. Skin protection is required when handling these materials: good extraction and ventilation facilities must be included within the processing area.

Injection molding of epoxides has a considerable advantage over other thermosets, as thick walled components can be produced using very short cycle times (i.e. less than 40 seconds). Molding of such components can take a tenth of the cycle time needed to produce the same part by compression molding. Components can be molded to very close tolerances as a result of the low variation in mold shrinkage and the absence of post molding shrinkage.

Screw designs for EMC have an L:D ratio of approximately 15 to 18:1, a compression ratio of 1 to 1.2:1 and, a flight depth of up to 4mm/0.157in. A back flow valve is not necessary but the screw tip should have an inclusive angle of 75 to 90 degrees. The preferred gap, beween the angle of the nozzle and the angle of the screw tip, (when the screw is in its most forward position) is 0.5mm/0.02in, or less, so as to allow for a nil cushion situation. It is usual to use a fully hardened, or ion nitrided screw, to counteract wear. However, wear to the screw is appreciably less than that encountered when processing PFs, MFs or UFs. One cause of screw wear, is the reverse flow of molten material along the barrel during mold filling. As EMC has a low viscosity at elevated temperatures (i.e. it will flow readily in the mold), the barrel temperatures can be set, so as to achieve a higher viscosity in the front section of the barrel, and this setting will resist any reverse flow. As with other thermosets, any increase in fill rate or injection pressure increases the risk of wear. In order to obtain a constant melt viscosity within the barrel, it is preferable to delay screw recovery so that screw rotation is completed just before the mold is opened. Screw surface speeds of up to 0.3m/s (1ft/s) are used.

EMC is being increasingly used for the production of moldings that include metallic inserts as, for example, EMC has excellent resistance to cracking. It is essential to ensure that the inserts are preheated prior to being placed in the mold. The preheating ensures that optimum cure is achieved throughout the molding, in particular around, or adjacent to, the metallic insert. Component failure while in service (e.g. cracking of material around inserts) has been traced back to unheated inserts being used during the processing operation.

11 Mold considerations

Materials contain a mold release (e.g. a metallic soap) which will build up to prevent adhesion to mold surfaces. Compression mold designs use similar design principles to those used for PFs. When designing molds for the long fiber filled grades, the positive and modified semi-positive types, are preferred so that excess resin does not flow from the mold during the pressurizing stage. The clearance between the punch and the die is 0.05 to 0.1mm/0.002 to 0.004in and, the angles on the loading chamber walls are approximately 0.5 to 2 degrees. The types of steels used for compression molds are based on nickel-chrome alloys (e.g. H.13): these are generally heat treated to 55 to 59RC and normally chrome plated (including ejector faces). Mold surfaces tend to keep clean as EMC does not stain.

Because of EMC's low demolding strength, extra ejection is required to prevent distortion, when the component is ejected from the mold. Heating capacity for molds is approximately 40 to 50 watts/Kg or 18 to 23 watts/lb. For transfer molds, a centralized pot type design, having equal runner lengths to each cavity, is preferred (particularly for encapsulation-type mold designs). The gates used for transfer molding should be rectangular in section (e.g. edge, film or fan-type gates). A typical formula used to determine gate sizes is:

$$\frac{\text{the volume of molding } (cm^3)}{2} = \text{Area of gate } (mm^2)$$

Venting of such molds should be provided, for example, at the end of the material flow path in the cavity or, via ejector pins. The width of the vents is approximately 6mm/0.236in and, the depth is approximately 0.05mm/0.002in.

For injection molds, the mold design must take into account the relatively high working temperatures (i.e. 180 to 220°C/356 to 392°F). Due to the low viscosity of EMC, the accuracy and quality of the mold manufacture must be good. Special attention should be given to the parting surfaces and, also to support within the mold so as to prevent material leakage at the parting line, as a result of mold plate deflection.

Due to the low molding shrinkage of EMC, undercuts may be needed to ensure that the molding remains on the correct half. However, because of its poor demolding strength, care must be taken not to use excessive undercuts as these can create distortion upon ejection. Typical sizes for undercuts are 0.2 to 0.3mm/0.008 to 0.012in deep. The gates used are of the pin, edge and film types; the land length should not be greater than 0.8mm/0.032in. Typical formulae used for EMC gate sizes are:

a. pin gates (area in mm^2) = 0.2 to 0.4 x molding volume (cm^3)

b. film gates (area in mm^2) = molding volume (cm^3) x 0.3 to 0.5 where the depth of the film gate = 0.5 to 1.5mm and the land length = 1.5 to 2mm.

As a result of the material's low viscosity, and insensitivity to high shear, small gates can be used; long flow lengths can also be achieved in the mold. Typical draft angles used are 1 to 2 degrees; venting and ejection techniques are similar to transfer design, hence vacuum venting is not required.

12 Ease of flow

These materials can be made to have very easy flow characteristics (they can be 3 to 5 times that of phenolics, aminoplastics and polyesters). Transfer molding grades are available which have siral flow lengths of up to 2540mm/100in; such materials are used in electronic applications for the potting of delicate components. Normally quoted as soft, medium and hard flow grades; as a rule of thumb, a hard flow grade ranges from 5 to 6in (127 to 152mm) spiral flow length, a medium flow grade would be 6 to 11in (152 to 280mm) and a soft flow grade would be 11in (280mm) and above. Long fiber filled grades are not sold as soft, hard or medium flow due to their stiffness of flow.

13 Shrinkage

Mold shrinkage for injection molding grades is 0.6 to 0.9% (0.006 to 0.009in/in or mm/mm), whereas the shrinkage for transfer and compression grades is 0.6 to 0.8%. Most compounds exhibit a very small variation in mold shrinkage (with nil post molding shrinkage), hence products requiring close tolerances can be readily molded. High GF filled grades will be lower, e.g. 0.2 to 0.4%.

14 Resistant to

A wide range of aqueous solutions and solvents. Being cross-linked the resin will not dissolve without decomposition, but will be swollen by liquids of similar solubility parameter to the cured resin. Chemical resistance is dependent on both the hardener and the resin as these, will determine the linkages formed, e.g. acidic hardeners form ester groups which will be less resistant to alkalis but the cured material has superior electrical properties. Such materials have good resistance to aqueous acids and salts and, to aliphatic hydrocarbons. Amine cured materials have, in addition, better alkali, inorganic salt and, organic chemical resistance.

15 Not resistant to

Polar organic solvents (e.g. methyl ethyl ketone (MEK), tetrahydrafuran (THF) and industrial methylated spirits (IMS); alkalis attack acid cured materials whereas amine-cured compounds have good resistance. Water/ozone mixtures can cause severe attack (also for many other plastics): such mixtures are now used in swimming pools.

16 Tests for cure or stability

Reflux a piece of the molding in boiling acetone (similar test to that used for PF – see ASTM-D-494). Moldings that do not exhibit blisters or swelling are usually properly cured: one cannot overcure EMC. Hardness may be measured, immediately upon demolding; this is used to check the degree of cure. Minimum Shore D hardness should be 55 to 57 but ideally should be 80% of the ultimate value, i.e. the value obtained if the molding was cured for a long time.

17 Detection or identification.

The unfilled resin has a density of 1.15 to $1.2\text{gcm}^{-3}/0.67$ to 0.69 oz/cu.in but as such materials are usually filled with inorganic fillers (e.g. silicas, metal silicates, glass fiber, etc.) the density is often $2\text{gcm}^{-3}/1.16\text{oz.in}^{-3}$. The moldings therefore sink readily in both water and in saturated magnesium chloride. The moldings are often large with abrupt changes in wall thickness as such changes can be readily accommodated with such materials. Epoxies can tolerate such treatment and still give dense porous-free moldings. Some epoxides, based upon Bisphenol A can be identified by immersing all or part of the molding into concentrated sulfuric acid (H_2SO_4); some EMC moldings change the color of the solution from clear to red.

When the material is placed in a flame, the component burns with a bright yellow flame. When removed, the flame extinguishes giving off dense white smoke with a pungent odor, and the edges of the burnt molding are swollen and black. The molding resists marking with sharp objects and when struck has a metallic ring. Components have a heavy, solid feel with only a little 'give' when flexed. Will brittle fracture if bent sufficiently. No obvious sign of shrinkage and moldings can be produced with a very high order of flatness and a good surface finish (resembles machined metal in this respect). Some moldings have incredible wall thicknesses, e.g. 20mm/0.787in. Components possess a dense solid coloration and, those cured with anhydrides, may be produced in light colors (white, beige, grey, etc.)

18 Component finishing.

For optimum properties, post-bake for 8 hours at 135°C/ 275°F, then raise the temperature at 10°C/18°F, per hour until a temperature of 215°C/419°F is reached. The improvement in properties may, however, be marginal if the molding conditions have been optimized. EMC is not usually decorated, but decoration can be carried out in the form of hot foiling, stencilling or silk screening. Quick setting paints are often used to fill in recessed lettering in order to highlight logos, etc.

19 Other comments

Epoxies are probably best known as good adhesives, for example, Araldite. Good bonding with such materials is achieved, partly because of the low curing shrinkage but also because of the polar hydroxy group; this helps initial wetting on polar substances. Hence the resins stick well to glass, stone, metal, etc. Because of EMCs chemical resistance, and ability to mold to close tolerances, they have replaced phenolic compounds for components used in textile machinery, where the service life has been doubled and, the need to machine the moldings, in order to obtain the desired tolerances, has been removed.

20 Typical components

Since their introduction (about 30 years ago), the most important applications of epoxy molding compounds have been in electrical engineering. This is because of the high tracking resistance, arc resistance and dielectric strength of epoxy components – even under extremely varying climatic conditions. Because of their good chemical resistance, epoxy compounds are used in chemical plant applications. Their outstanding dimensional stability, and lack of internal stresses, allows EMCs to be used where the molded components contain metallic inserts (e.g. automotive distributor caps).

Typical uses vary from spark plugs connectors, thermostat control mechanisms, pump assemblies and, circuit breaker assemblies. Epoxides based upon acid anhydrides can produce light colors, which has led to use in laboratory furniture (bench tops, basins, etc.). However, such items are manufactured by the casting technique and not by a molding operation.

21 Unit Price

The price relative to PF (General Purpose Grade) is as follows:

Injection molding grade	2.83
Filled injection molding grade	4.16
Long fiber compression molding grade	5.14

The price relative to GPPF can also be found in Table 6.

SUGGESTED PROCESSING CONDITIONS FOR EMC

Process	Setting	Unit	Value	
			Long fiber filled	Granular material
Compression	Preheat temperature	°C	80 to 100	80 to 100
	Tool temperature	°C	150 to 180	150 to 200
	Pressure	MNm^{-2}	20 to 50	5 to 20
	Curing time	s	120	80
Transfer	Tool temperature	°C	150 to 180	150 to 200
	Transfer pressure	MNm^{-2}	20 to 50	10 to 50
	Curing time	s	150	50
Injection	Cylinder temperature	°C	-	50 to 70
	Nozzle temperature	°C	-	70 to 90
	Melt temperature	°C	-	80 to 100
	Screw temperature	°C	-	40 to 70
	Tool temperature	°C	-	180 to 220
	Injection pressure	MNm^{-2}	-	80 to 180
	Back pressure	MNm^{-2}	-	2 to 12
	Curing time	s	-	20
	Injection time	s	-	3 to 15

Process	Setting	Unit	Value	
			Long fiber filled	Granular material
Compression	Preheat temperature	°F	176 to 212	176 to 212
	Tool temperature	°F	302 to 356	302 to 392
	Pressure	psi	2,900 to 7,250	-
	Pressure	psi	-	725 to 2,900
	Curing time	s	120	80
Transfer	Tool temperature	°F	302 to 356	302 to 392
	Transfer pressure	psi	2,900 to 7,250	-
	Transfer pressure	psi	-	1,450 to 7,250
	Curing time	s	150	50
Injection	Cylinder temperature	°F	-	122 to 158
	Nozzle temperature	°F	-	158 to 194
	Melt temperature	°F	-	176 to 212
	Screw temperature	°F	-	104 to 158
	Tool temperature	°F	-	356 to 428
	Injection pressure	psi	-	11,610 to 21,600
	Curing time	s	-	20
	Injection time	s	-	3 to 15

GRANULAR POLYESTER MOLDING COMPOUNDS.

1 Common name(s)

Granular polyester compound; granular polyester molding compound.

2 Abbreviations

GPMC; GMC; NMC.

3 Alternative names

Granular molding compounds (GMC); nodular molding compound (NMC); granular polyesters; alkyds; polyester alkyds.

4 Some suppliers	5 Trade names or trade marks
Freeman Chemicals	Freeflo
Scott Bader	Impel
Sterling Moulding Materials	Rutaform Polyester
Sterling Moulding Materials	Bakelite Polyester Alkyd
Resin suppliers.	
BASF	Palatal
Bayer	Leguval
Fiberglass	Dion and Vibrin
Mia Chemicals	Miapol
Scott Bader	Crystic
SIR	Sirester
PMC compounds.	
Ashland Chemical	Aropol and Hectron
Chromos - Ro - Polimeri	Esterform
Dr. Illing	Illandur
Haysite Industrial Plastics	Haysite
Isola Werke	Durapol
Orkem	Norsomix
Pheonix	Keripol
Resart	Resatherm
Toyobo	Vyloglas
Ubbink Nederland	Uromix

6 Material properties

The term "ester" is given to the reaction product of an acid and an alcohol. A polyester is a reaction product which contains many ester groups and an unsaturated polyester resin (UP) is a reaction product which is capable of further reaction via what is known as a double bond (to ease application, such resins are usually dissolved in styrene monomer - see DMC and SMC). Many years ago the term alkyd was also applied to ester-type materials; the "al" came from the alcohol and the "kyd" from the acid. The term "glyptal" was also used for the products (polyesters) made from glycerol and phthalic anhydride.

Although UP resins are normally made from phthalic anhydride, maleic anhydride and a glycol such as propylene glycol, many other starting materials may be used. These may be chosen to give a solid resin after reaction: such a resin may then be blended with monomers, other than styrene, to give solid systems. When a non-volatile monomer (such as diallyl phthalate or diallyl isophthalate) is used, the molding composition which results after fillers, catalysts, etc. have been added, is sometimes known as an alkyd, a polyester alkyd or, an alkyd molding compound (AMC).

The term alkyd is not nowadays widely used; the term granular polyester molding compound is preferred. Use of the initials PMC, standing for pelletized molding compound, can lead to confusion as the same initials are now being used as a generic term for all molding compounds based on unsaturated polyesters (e.g. SMC, DMC) and, for powder mold coating.

A wide range of grades is possible as even if the resin and monomer are kept constant, the filler type, size and concentration may be varied. Like DMC, this type of material usually contains both mineral and fibrous fillers - the fibrous filler is usually glass fiber (GF). Unlike DMC, GMC can be molded into complex shapes without causing resin/filler separation in deep sections, such as ribs and bosses.

7 Forms supplied

Supplied as pellets. Pellets of this material resemble thermoplastic pellets and are sold like them packed into standard sized bags, for example, 25kg/55.1lb bags. Such pellets can be hopper fed to the molding machine. GMC may be automatically fed by the use of hopper loaders from large storage bins.

8 Materials handling, packaging and storage

Material should be stored in a cool, dry room (15 to 20°C/59 to 68°F). Avoid storing bags in direct sunlight; GMC usually has a storage life in excess of 6 months if these conditions are observed. Dry, or granular PMC, usually has a longer storage life than styrene-based materials; granular PMC is usually styrene free. Such molding compounds are often crosslinked with diallyl phthalate. Some black GMC is not so tolerant and should be used within 3 to 6 months from the date of manufacture.

9 Color possibilities

Wide color range is available. The material is already colored (as a rule cadmium based colors are more expensive). Color matching can be more difficult than for DMC/SMC materials.

10 Molding conditions

GMC can be compression molded successfully. However, granular boundaries can be apparent on moldings, which tend to spoil their appearance. The boundary effect is dependant upon the shape of the article to be produced and, whether the material can flow in the mold. When the material is merely compressed, and does not really flow, then boundary problems will be encountered. In general high compressive forces (tonnages) are usually adopted.

GMC can also be transfer and injection molded. The design of transfer pot and feed system is similar to that used for DMC.

With injection molding, the screw design is not as critical as for DMC; the actual design used is similar to that for phenolics. There is no need for a back flow valve assembly, as GMC is not so fluid along the barrel length. Although it is not at all recommended, the material has been molded using a GP screw design for thermoplastic materials. The material is non-hygroscopic, so predrying is not necessary. It has good barrel life at molding temperatures, so the risk of curing in the barrel is low. As the fibers in GMC are very short they are not susceptible to further damage upon molding (as with DMC), which leads to less critical plasticizing conditions. Because the parts made from GMC are normally small, cooling jigs are not always required. Accurate barrel temperature control is required as GMC has a relatively sharp melting point. Material based on crystalline resins changes from solid to liquid within 4 to 5°C (7.2 to 9°F), whereas amorphous type resins have a wider softening range (10 to 15°C/18 to 27°F).

11 Mold considerations

Unlike DMC, vacuum extraction of the mold (to prevent porosity) is not widely used. If the components contain air traps, such as deep sections, venting may be necessary.

Because of GMC's easier flow characteristics, smaller runner systems can be tolerated than for DMC/SMC as fiber breakage is not as much of a problem. Use of smaller runners (4 to 6mm/0.157 to 0.236in diameter) also increases the pressure that needs to be applied to the material thus improving material consolidation. A range of gates can be used (edge, tab, film, sprue, etc.) and, in general, gate sizes tend to be smaller than those used for DMC/SMC and phenolic. The actual dimensions of the gate are related to the size of the product. For example, to produce a component having an area of $484cm^2$ use a film gate which is 1.5mm deep x 51mm long: or, to produce a component having an area of $75in^2$ use a film gate which is 0.059in deep x 2in long.

As the majority of GMCs are GF filled, any abrasion in the gate area, or in the mold surface, would be similar to that experienced with a DMC/SMC of the same glass content. It therefore follows that the lower the glass content, the lower is the surface abrasion is encountered. Demolding is similar to DMC, where moldings do tend to be soft on ejection. However, distortion upon ejection can be avoided in correctly designed molds (see DMC).

12 Ease of flow

MFR at $105°C/221°F$ is 8g/600s. Good flow properties mean that complex shapes can be molded at high speeds with low pressures.

13 Shrinkage

Shrinkage varies from 0.2 to 0.8%. Flame retardant grades are in the same range. There is virtually no after shrinkage.

14 Resistant to

High temperatures – GMC can withstand up to $220°C/428°F$, whereas DMC would not be recommended above $160°C/320°F$. The chemical resistance is similar to DMC.

15 Not resistant to

Boiling water, more affected than allyls on exposure to humidity or boiling water. As with DMC, acidic environments can cause problems due to the widespread usage of calcium carbonate fillers in this type of material.

16 Tests for cure or stability

The Barcol hardness (Barcol 934) will confirm the material is fully cured on cooling.

17 Detection or identification

The molded material can have a high density - around $2gcm^{-3}/1.16oz$ in^{-3} – and components will therefore sink rapidly in water. The apparent density of pellets is approximately $0.85gcm^{-3}/0.49oz$ in^{-3}. This type of material is available in a wide range of colors, has good heat resistance and resists burning - oxygen index is about 50% (but can very from 25 to 66% dependant upon formulation) and, the material is rated as V-0 by UL94. The appearance of moldings ranges from a satin finish to a high gloss finish: such a finish is usually free from flow marks. In general, the surface finish obtained is far better than can be achieved from DMC/SMC. The surface cannot be scratched with a fingernail but marks relatively easily with a knife: large chips can be easily cut with a knife. When placed in a flame, it is slow to ignite and a small yellow flame forms with cracking noises being heard. When removed from the flame, the material self-extinguishes, with white smoke forming giving off an acrid odor. Burnt edges of the molding are swollen and black. When struck with a metal object the material is relatively brittle and the hole produced tends to be larger on the side opposite the striking force (a crater-like hole is produced).

18 Component finishing or decoration

The finishing and deflashing techniques are similar to those used for DMC. Due to the better control of flash, and smaller gate scars, finishing costs for GMC are less than those for DMC. Due to the dry granular nature of the pellets, the resultant moldings are less prone to contamination than DMC/SMC materials.

19 Other comments

GMC has been successful in replacing other thermosets such as phenolics for certain products as a result of faster production and superior properties. The capital expenditure involved in replacing phenolics is minimal due to its similar processing and shrinkage characteristics.

20 Typical components

GMC materials are used for the manufacture of automotive accessories such as distributor caps, ignition coil caps and lamp sockets. Also used for domestic appliance applications such as iron handles, toaster or grill housings and, switches for washing machines and dishwashers.

The good mechanical stability and excellent insulating qualities, are properties which allow the successful use in components such as connector boxes for telecommunication systems and, casings for explosion proof switches.

A major development is in the use of GMC for microwave cookware. As these materials can withstand up to $220°\text{C}/428°\text{F}$, microwave trays can be hot filled, frozen and reheated without apparent damage. As the materials have good transparency to microwaves, and can withstand cooking conditions, they are widely used for microwave cookware - both re-usable and disposable. DMC dishes are glossier but must be post-baked to remove traces of styrene.

Allyl and GMC materials have, broadly speaking, similar properties, use, and employ similar processing conditions. Both are dry, granular solids. GMCs are, however, weaker mechanically than allyl materials, have a lower resistance to cracking around inserts, do not maintain their electrical properties so well under extreme conditions, but are cheaper.

GMC materials do not have the same strength properties as DMC (the GF is shorter) but have the advantages of being dry granular solids – they look and handle like pellets of a thermoplastic material. Because they are pelleted, there is no need for an expensive stuffer unit (as is necessary for DMC). As each DMC machine needs a stuffer unit, this is a tremendous saving.

21 Unit Price.

The price relative to PF (General Purpose Grade) is 1.41. The price relative to GPPF can also be found in Table 6.

SUGGESTED PROCESSING CONDITIONS FOR GMC

Process	Setting	Unit	Value
Compression	Preheat temperature	$^\circ$C	70 to 80
	Tool temperature	$^\circ$C	150 to 190
	Pressure	MNm^{-2}	5 to 10
	Curing time	s	10 to 20s/mm
Transfer	Tool temperature	$^\circ$C	150 to 190
	Transfer pressure	MNm^{-2}	10 to 80
	Curing time	s	10 to 60 more or less independent of wall section.
Injection	Cylinder temperature	$^\circ$C	30 to 80
	Nozzle temperature	$^\circ$C	60 to 100
	Melt temperature	$^\circ$C	100 to 140
	Screw temperature	$^\circ$C	30 to 60
	Tool temperature	$^\circ$C	150 to 190
	Injection pressure	MNm^{-2}	50 to 150
	Back pressure	MNm^{-2}	4 to 15
	Curing time	s	10 to 100 more or less independent of wall section.

Process	Setting	Unit	Value
Compression	Preheat temperature	$^\circ$F	158 to 176
	Tool temperature	$^\circ$F	302 to 374
	Pressure	psi	725 to 1,450
	Curing time	s	0.5 to 1s/0.002in
Transfer	Tool temperature	$^\circ$F	302 to 374
	Transfer pressure	psi	1,450 to 11,600
	Curing time	s	10 to 60 more or less independent of wall section.
Injection	Cylinder temperature	$^\circ$F	86 to 176
	Nozzle temperature	$^\circ$F	140 to 212
	Melt temperature	$^\circ$F	212 to 284
	Screw temperature	$^\circ$F	86 to 140
	Tool temperature	$^\circ$F	302 to 374
	Injection pressure	psi	7,250 to 21,750
	Back pressure	psi	580 to 2,175
	Curing time	s	10 to 100 more or less independent of wall section.

MELAMINE FORMALDEHYDE

1 Common Name

Melamine formaldehyde; melamine formaldehyde molding material.

2 Abbreviation

MF.

3 Alternative Names

Melamine resin molding material or compound. MF, like UF, is an aminoplastic that is, the molding powders are based on an amino-formaldehyde resin.

4 Some Suppliers	5 Trade Names or Trade Marks
American Cyanamid	Cymel
BIP	Melmex
Ciba Geigy	Melopas
Huls (Dynamit Nobel)	Ultraplas
Perstorp	Isomin
Perstorp Ferguson	Melamine molding compound
SPREA	Melsprea
Sterling Moulding Materials	Bakelite-Melamine.
Sterling Moulding Materials	Rutaform

6 Material Properties

Melamine may be made from urea and like urea it contains amine ($-NH_2$) groups. These react with formaldehyde to form a resin which is combined with fillers, hardeners, pigments, etc. to make a molding powder. As with colorless UF resins, it is possible to produce molding powders of virtually any color when white alpha cellulose, or a paper filler, is used. Moldings are bright and attractive but unfortunately they are more expensive than UF or PF. They have lower water absorption than UFs, are less flammable, maintain their electrical properties better in conditions of high humidity (particularly when mineral filled), are harder and, resist staining and heat to a greater degree.

Like UFs, they show after-shrinkage if exposed continuously to high temperatures but maximum, continuous, end-use temperature is much higher than UF at 130°C or 266°F. Like PFs, they are more tolerant of changes in molding conditions than UF materials. They are available in a very wide color range and possess better scratch and tracking resistance than PF materials; they are not so impact resistant as PF but have good abrasion resistance. Better electrical properties are obtained from melamine-phenolics. Where moldings are to be used alternatively in wet and dry conditions, the use of a plasticized grade is suggested: benzamide and p-toluene sulfonamide have been used as plasticizers. New improved injection molding grades have excellent color retention when exposed to heat and sunlight and, can now be considered for use at continuous temperatures of up to 130°C, or 266°F, in wet and dry conditions. Due to the low viscosity, of this type of grade, the manufacture of large and complex articles is possible. Apart from better processability, such grades give high gloss, improved impact resistance and, a lower post-molding distortion.

7 Forms Supplied

Usually supplied as granular powders which have a bulk density of about 2:1. Particle size, and size distribution, is adjusted during manufacture so that the material may be loaded either, by weight or by volume or, it may be pelleted and used with or without preheating. Grades are chosen to suit the process and application, for example compression, transfer and injection molding. Special grades are available for the compression molding of plates; the use of such grades gives an even, hard, scratch resistant surface. These grades are usually in the form of fine powder - rather than a granular material.

8 Materials Handling, Packaging and Storage

Material is supplied in 25kg/55.1lb bags which should be stored in a cool, dry place (lower than 20°C or 68°F) and protected against moisture. If this is done the shelf life will be at least one year. Ensure that the materials handling system does not cause molding powder segregation (by preferential lifting of smaller particle sizes). This could cause shot to shot variation when injection molding.

9 Color Possibilities

There is an unlimited choice of color, with a wide range of colors offered by manufacturers. MF gives very even pigmentation and may be color coded according to the RAL system. MPF materials are also available in white and pastel shades; color stability is adequate for many uses.

10 Molding Conditions

When injection molding, the highest mold temperatures are for the thinnest sections (below 3mm or 0.118in). Thicker sections require lower temperatures, around 155°C or 311°F, but longer cure times. Keep the screw speed low, (about 50rpm) and, fill the mold as quickly as possible – consistent with avoiding gas trapping. Be very careful to avoid overpacking; use up to 30% of the injection pressure as packing pressure and bring this in when the screw is about 10mm, or 0.394in, away from bottom. At this point, it is possible to arrange for mold breathing to occur: although MF is not as gassy as UF, good venting and/or breathing is still necessary. When processing melamine phenolics, screw surface speeds of up to 0.3m/s, approximately 1ft/s, are permissible for plasticizing.

In compression molding, MF is usually run hotter than UF, at around 160°C or 320°F. At this temperature a 2mm, or 0.079in, thick component will need a cure of 30 seconds; a 4mm, 0.157in, section will need 60 seconds. To prevent cracking around large inserts, warm them to the mold temperature before use. If the material is preheated by high frequency preheating (HF) then, the equipment must be powerful enough to raise the temperature to 100°C, or 212°F, in less than 60 seconds to avoid molding problems. Preheating improves appearance, electrical properties, dimensional stability and output. Compared to UF materials, MF is normally pelleted and preheated prior to introduction into the mold. Materials handling can be improved by producing a heated slug of material on an in-line screw unit – this saves pelleting and preheating.

With melamine-phenolic materials (MPF), preheating improves the quality of finished molding and, in particular, maintains post molding shrinkage within narrow limits. Post molding shrinkage can be a problem with MF and can result in component failure. The reduced post molding shrinkage of this class of materials, is a significant advantage over MFs. For transfer molding of melamine phenolics, preheating is always advisable.

When processing any plastics material which is based on formaldehyde, care must be taken and the fumes evolved treated as harmful: good ventilation of the work-place is essential.

11 Mold Considerations

In many cases the principles used for mold design are similar to those used for UF materials (see UF). However, with low viscosity injection grades of MF, the size of runners and gates can be reduced due to the easier flow and mold filling characteristics: typical runner sizes are 3 to 4mm or 0.118 to 0.159in.

Excessive distortion, of the end product, can be substantially reduced by using a film gate: this also enables the cavity to be filled more uniformly and much reduces orientation (and the resultant anisotropy).

When using conventional venting techniques, for injection molding, the size of vent used for low viscosity resins should be smaller than for standard grades as pressure loss in the cavity may be experienced; such losses will result in the moldings possessing sink marks or porosity. Typical vent sizes for low viscosity grades are 0.04 to 0.07mm (0.0015 to 0.003in). Vacuum venting is the most efficient form of venting when injection molding MF.

12 Ease of Flow

MF materials can be of comparatively low viscosity. Such flow properties, permit the injection molding of large complex handles and bowls which gives design freedom.

13 Shrinkage

About 0.7% or, 0.007in/in or, 0.007mm/mm. Like UFs, after-shrinkage will occur if the molding is held at high temperatures. Such shrinkage is usually identified by testing at 110°C (230°F) for 48 hours; after this time, the MF molding will shrink up to 1.2%

14 Resistant to

MF moldings are resistant to fuels, oils, greases, common organic liquids and organic solvents - such as acetone and alcohol. Also resistant to cold dilute acids and alkalis. MF is more chemically resistant and more stain resistant than UF, that is with better resistance to weak acids, alkalis and water. UF moldings are attacked by boiling water; fully cured MFs are more resistant, being attacked only by concentrated acids and alkalis or, by hot dilute acids.

15 Not Resistant to

Strong acids and alkalis and, to hot dilute acids. Continuous exposure to high temperatures can also cause problems. MF is better than UF in this respect – an MF molding can withstand 130°C (266°F) continuous maximum working temperature. A UF molding can stand only 75°C (167°F) continuous maximum working temperature. Both these aminoplastics suffer from after-shrinkage if held at high temperature for too long.

16 Tests for Cure or Stability

An MF molding requires a more severe cure test than that used for a UF molding. Boil a part of the molding gently in an 0.01% aqueous solution of Rhodamine B for 10 minutes, remove with tongs and, inspect the molding for staining. Ignore the cut edges, and flash lines, when the molding is inspected for staining. If the molding is properly cured, only slight staining will be seen and, the moldings will have a good balance of properties.

To test for internal stresses, stove the moldings for 48 hours at 60°C, or 140°F, and inspect for cracks: good quality moldings will easily withstand this treatment.

17 Detection or Identification

Moldings are opaque and can be any color from white to black. They may also have surface decoration, for example, imparted by the use of molded-in foils. The moldings are difficult to cut with a knife; small chips are cut with difficulty, and the molding cannot be marked with a fingernail. White scratches are formed when the surface is marked with a knife (due to the appearance of white filler). Moldings will crack when struck sharply with a metallic object.

When a molding is placed in a flame, it is not initially easy to ignite but then, a yellow-orange flame appears and cracking sounds are heard; the bulk of the burnt area is dark grey with white to yellow edges. Burned edges glow in the flame. There is a bitter acrid odor when the molding is removed from the flame (smells of burnt fish) and, dense white smoke is often seen: trace amounts of cyanides may also be present so, take great care when doing heating tests. MFs can withstand higher temperatures before decomposition, or degradation, than can UFs.

As the density is about 1.5 gcm^{-3}, 24.58g/in^3 or 0.87oz/in^3, the moldings sink readily in both water and, in saturated magnesium chloride solution. MF behaves similarly to UF in simple tests: on heating in hot dye solutions, UF moldings stains easily, while MF moldings resist such staining (even in acidic dye solutions).

18 Component Finishing or Decoration

Multi-color decorative foils (printed alpha-cellulose paper

impregnated with uncured MF resin) can be introduced into compression molds containing minimally-cured MF moldings before completing the cure. These relatively cheap foils (such as Ornamin) enable personalized moldings, for a number of customers, to be produced from one tool. Although more easily produced in flat or smooth profile molds, grades of foil are available for sharp and rectangular edges.

The excellent color range also enables two-tone, compression moldings to be made using double-punch tools; for instance, cups and mugs with a white inside and, a colored exterior.

Sometimes paints are used to highlight recessed lettering and, printing has been used for lettering or for general decoration over a base color.

19 Other Comments

MF resins are widely used in laminates as by their use it is possible to produce a very wide range of attractive, and durable, patterned laminates – the pattern printed on the paper base shows through the colorless resin. The laminate core is made of cheaper, PF-coated paper. Such laminates have excellent light stability. MF resins are also used for textile treatments and for paint manufacture.

20 Typical Components

Because molded components can be free from odor (they do not influence the flavor of food) are bright, attractive, scratch and stain resistant, they are widely used in dinnerware and kitchen utensils. Such items can be made more attractive by molding-in foils which carry a legend or picture.

Formaldehyde is very toxic, and carcinogenic, and so free formaldehyde is the major cause for concern when dealing with the use of aminoplastics which come into contact with food. The toxicity of free melamine and of free urea is low in comparison with that of formaldehyde. However, as melamine is slightly soluble in water and does have a diuretic effect on animals, care should be exercised when dealing with foodstuffs. Standards exist, for example, BS 2782, which specify how free formaldehyde is measured in MF resins.

MF is also used for handles, knobs, household appliances and electrically insulating parts – particularly where these must withstand high temperatures. It can withstand repeated cycling from 150°C (302°F) into cold water. Compared to other thermosets, MFs are expensive and are only selected where appearance is of prime importance. Where improved impact strength is required, glass fiber is used as a filler – it gives high mechanical strength and heat resistance; the moldings possess good arc and track resistance and, burn with difficulty. Examples of moldings which utilize these properties, include base plates for circuit breakers and switches. The filling of MF with inorganic fillers, such as glass, can give hard, heat and mar resistant moldings. Such materials can be made very fire resistant, they burn with great difficulty, and are used in applications which demand this property, for example, on board ships.

21 Unit Price

The unit price relative to PF (General Purpose Grade) is as follows:

Compression molding grade (white)	1.69
Compression molding grade (colors)	1.85
Injection molding grade (white)	1.73
Injection molding grade (colors)	1.87

The price relative to GPPF can be also found in Table 6.

SUGGESTED PROCESSING CONDITIONS FOR MF AND FOR MPF.

Process	Setting	Unit	Value	
			MF	**MPF**
Compression	Preheat temperature	°C	<100	<100
	Tool temperature	°C	140 to 180	150 to 175
	Pressure	MNm^{-2}	15 to 60	20 to 60
	Curing time	s	40 to 210	20 to 40
Transfer	Tool temperature	°C	150 to 170	140 to 170
	Transfer pressure	MNm^{-2}	60 to 120	80 to 150
	Curing time	s	30 to 120	60 to 120
Injection	Cylinder temperature	°C	85 to 105	70 to 80
	Nozzle temperature	°C	95 to 120	110 to 120
	Melt temperature	°C	120 to 140	120 to 140
	Screw temperature	°C	65 to 80	65 to 80
	Tool temperature	°C	145 to 180	165 to 180
	Injection pressure	MNm^{-2}	100 to 248	100 to 175
	Back pressure	MNm^{-2}	<1 (eg0.5)	8 to 12
	Curing time	s	15 to 80	25 to 50

Process	Setting	Unit	Value	
			MF	**MPF**
Compression	Preheat temperature	°F	<212	<212
	Tool temperature	°F	284 to 356	302 to 355
	Pressure	psi	>2,200	>2,700
	Curing time	s	40 to 210	20 to 40
Transfer	Tool temperature	°F	302 to 338	284 to 338
	Transfer pressure	psi	>7,250	>11,300
	Curing time	s	30 to 120	60 to 120
Injection	Cylinder temperature	°F	185 to 221	158 to 176
	Nozzle temperature	°F	203 to 248	230 to 248
	Melt temperature	°F	248 to 284	248 to 284
	Screw temperature	°F	149 to 176	149 to 176
	Tool temperature	°F	293 to 356	329 to 356
	Injection pressure	psi	14,500 to 36,000	14,500 to 25,500
	Curing time	s	15 to 80	25 to 50

PHENOL-FORMALDEHYDE.

1 Common name(s).

Phenolics; phenolic molding materials.

2 Abbreviation

PF.

3 Alternative names

Bakelite; phenoplast; phenolic resin molding compounds; novolak resin molding compounds..

4 Some suppliers | 5 Trade names or trade marks

4 Some suppliers	5 Trade names or trade marks
Bakelite	Bakelite
BP Chemicals	Cellobond BP
Canadian Oxy Chemicals	Durez
Chemiplastica	Fenochem
Chromos - Ro- Polimeri	Fenoform
General Electric	Genal
Kingsley & Keith	Sirfen
Perstorp	Nestorite
Perstorp Ferguson	Phenolic moulding compounds
Plastics Engineering	Plenco
Reichold	Dynaset
Rhone Poulenc	Progilite
Rogers Corp	Phenolic moulding compounds
SIR	Sirfen
SPREA	Moldsite
Sterling Moulding	Bakelite and Sternite
T.B.A.	Durestos.
Toshiba Chemical Products	Tecolite

6 Material properties

Phenolic resins are based on the resinous material formed when a phenol-type material is reacted (or condensed) with an aldehyde. Phenol itself is widely used as the resins produced have good mechanical strength and cure quickly. Cresols may be used for more acid-resistant products and the use of phenol/cresol mixtures lowers the cost and controls the flow in processing, for example, in compression molding, the use of 20% cresol may be employed. Where improved alkali resistance is required, xylenols may be used. Formaldehyde is far and away the most widely used aldehyde.

The unreinforced resin is very brittle and requires extensive modification with fillers to produce useful products. A very wide range of properties can be obtained from PFs because of their compatibility with a variety of reinforcements and fillers, so the properties are very dependent on filler used. Woodflour gives reasonable properties at an acceptable cost, and so woodflour-filled PF is regarded as a general purpose (GP) material. The use of a more fibrous, organic filler (cellulose fibers) gives improved toughness and impact strength.

Modification with mineral fillers yields increased rigidity, improved dimensional and, thermal stability; results in a higher UL index of use, lower water absorption and thermal expansion. Glass fiber (GF) addition, can improve the dimensional stability and rigidity even more: the UL index of use can reach $180°C/356°F$. Because of density differences, organic fillers are used in a weight ratio of 1:1 with the resin and inorganic fillers (mineral and glass) used in a ratio of 1.5:1.

Moldings are glossy, opaque and dark-colored but have the tremendous advantage of being inherently flame retardant, low smoke systems which do not require halogen or phosphorus-based additives. They are stiff, hard, have low elongations and possess good creep resistance..

The materials are commonly supplied in powder form and have a useful combination of low cost, ease and versatility of molding, temperature resistance, solvent and chemical resistance and, good electrical insulation properties. They have better water resistance than melamine formaldehyde and possess a more stable melt rheology than aminoplastics, i.e. they are not so temperature dependent. Tracking resistance under conditions of high humidity is poor and their impact resistance is not very good. Electrical properties, especially tracking resistance is inferior to aminoplastics. Melamine phenolics (MPF) have superior electrical properties to phenolics and have a wider color range: they are used in decorative and electrical application - that is, in areas which are beyond PFs. The reduced post molding shrinkage of this class of materials, is a significant advantage over MFs.

7 Forms supplied

Depends on the filler used. Woodflour and mineral fillers give powders (resembling coarsely ground coffee). Glass fiber (GF) grades are supplied in nodules for injection molding: a loose, fluffy, rag-like material may be supplied for injection molding. Fabric filled grades may resemble shredded rag but injection grades can be supplied in nodular form.

Powder grades are available in differing granulations to suit the process. For example, fine powders are used in compression molding where limited flow is envisaged but a high gloss is required. Fine-free grades are preferred in automatic compression molding, so as to avoid premature melting of the fines in the hopper. When injection molding, choose a granulation that is large enough to flow freely from the hopper and gives consistent pick-up on the screw.

8 Materials handling; packaging and storage

Normally supplied in strong paper, or plastic, sacks which hold 25kg/55.1lb, or 50lb/22.6lb, of material. Store the material in a cool (15 to 20°C/59 to 68°F), dry place and keep the bags sealed until required. Do not put material which escaped from a damaged bag back into the sack. With some grades, the storage stability is limited and due account must be taken of this by adopting a Fi-Fo policy and, therefore, strict control over residence time in the stores. Transfer sealed bags to the molding shop approximately 12 hours before use so as to allow temperature equilibration. If the bag is marked with an "Andrews Cross" then the material contains more than 1% of free phenol and should be treated accordingly. Most grades contain less than 1%.

9 Color possibilities

Because of the formation of complex molecules (quinone methides), PFs are naturally dark-colored materials which darken even more in sunlight or, on warming. As the unmodified material is virtually useless (for example, excessive shrinkage on curing and poor physical properties) the resin is extensively modified by the use of fillers: the presence of fillers makes the molding compounds opaque. The color range is therefore very limited, with black, dark brown, dark red and dark green being the most common colors. Recently comparatively pale shades of red have been produced. One way of overcoming the limited color range of PF materials, is by painting them after molding; the appearance of PF moldings may be changed dramatically by painting. In mold coating may be another.

10 Molding conditions

Injection barrel temperatures usually lie between 65 to 90°C/149 to 194°F. Set to the lowest temperatures possible and commence molding: raise temperatures in 5°C/9°F increments until signs of excessive overheating appear.

(These include gassing when air shotting, erratic screw movements, an increase in injection time and, inconsistent cavity filling). Reduce barrel temperature until cycles become consistent. When molding small shots, keep the back zone temperatures low at approximately 65°C/149°F. Screw speeds can be up to 100 rpm., but keep as low as possible. Only use sufficient back pressure to give consistent shot weight. Initially use low mold temperatures but increase until defects are observed and then reduce slightly. If the shot size becomes erratic, inspect the screw for wear (this first shows up on the front face of the first flight) and rebuild as soon as possible. Consistent changeover from first stage pressure to holding pressure is essential to control packing and flashing: this setting must be capable of being precisely set and held.

When compression molding, remember that preheating decreases mold shrinkage, increases post-mold dimensional stability, and improves the dielectric strength by up to 300%. Compression molding preserves the length of glass fibers, minimizes their orientation and produces parts with high flexural and tensile strength.

When processing any plastics material which is based on formaldehyde, care must be taken and the fumes evolved treated as harmful: good ventilation of the work-place is essential.

11 Mold considerations

High quality nickel chrome steels should be used, suitably hardened and chrome plated. Maintain the plating and polish as wear occurs so as to ensure lack of mold staining and ease of ejection. When injection molding, tool temperatures of 160 to 180°C/320 to 356°F are usual. Vents may be tapered with a narrow end at the tool face: the vent then widens as the cavity is left behind. For example 3mm x 0.05mm initially, widening to 5 x 3mm or, 0.118 x 0.02in initially, widening to 0.197 x 0.118in. Keep the sprue short, generous and well tapered (up to 5 degrees). Carry the gates on replaceable pads so that they may be easily replaced as wear occurs; ion implantation and/or titanium nitriding should help minimize wear.

Bulk factor of general purpose compression grades is approximately 2.3. High impact grades can reach 3.6. Powders are generally easy to pelletize (provided they have the right particle size distribution) but fabric filled grades are not.

12 Ease of flow

Available in a range of flows - soft (easy), medium and hard (stiff). In general, PFs are hard, stiff materials which require high pressure to get smooth mold filling. PFs have a more stable melt rheology than aminoplastics in that they are not so temperature dependent. The flow, and cure rate, of a molding material is determined by the amount of resin, the degree of condensation of the resin and, the residual volatile content, for example, the moisture content. If the resin is only lightly condensed during manufacture then, the resulting molding material will flow easily but will require comparatively long cure times: the moldings will exhibit high shrinkage and will not exhibit the best mechanical properties. Increasing the degree of condensation will worsen the flow but improve the properties of the moldings. About 1% of a plasticizer, or flow promoter such as dibutyl phthalate, may be incorporated in some low shrinkage grades to get adequate flow.

13 Shrinkage

For GP compression molding grades, it is about 0.8% that is 0.008mm/mm or in/in: that of glass-filled grades may be reduced to 0.3%. The shrinkage of acid-resistant and minimum odor grades will be higher than GP. Injection grades (GP) will range from 1 to 1.7% but impact and heat resistant grades could be as low as 0.8%.

14 Resistant to

Dilute acids (not all grades), alcohols, aromatic and chlorinated hydrocarbons, petrol, greases, fats and oils. PF moldings are resistant to organic chemicals, oils and fats even at elevated temperatures. Mineral-filled grades may also be detergent resistant. PF moldings retain their properties on ageing and have good dimensional stability over a wide temperature range.

15 Not resistant to

Strong mineral acids and alkalis: however, some grades are not so resistant and may even be attacked by dilute acids, alkalis, ketones and detergents. Those with organic fillers, and therefore higher water absorption, are more strongly attacked, on long exposures, than those grades filled with inorganic fillers. All grades should resist short term exposure to water. Sunlight will cause general darkening.

16 Tests for cure or stability

On ejection from the mold, the moldings are still capable of being distorted (as cure is not taken to the ultimate for economic reasons). To stop distortion, larger components may be placed on jigs for a few minutes while they cool. In some cases, leaving the components on the feed system will stop distortion as the runner acts as a stiffening bar. Amount of acetone-extractable material is used to assess extent of cure: hardness could also be used. Adequately cured moldings should show no significant damage after being boiled for approximately 30 minutes in water or, in dye solutions (see MF and UF).

17 Detection or identification

Unless painted, moldings are dark colored, for example, dark brown or black: the colors are usually dense and very opaque. The moldings feel stiff and hard. Resin content varies with grade but is usually 45 to 55% (high resin contents give longer flow and a better surface finish). The SG depends on filler (or filler combination) used: the density ranges from 1.3 to $1.8gcm^{-3}$ so moldings sink rapidly in water: the high SG materials will also sink rapidly in saturated magnesium chloride. The SG is most commonly about 1.4 and the moldings feel solid and heavy; they are also difficult to cut. When scraped, filed or sawn there is an odor of phenol (carbolic acid). Moldings soften slightly at high temperatures (around $130°C/266°F$) and, only burn with difficulty. The material cracks when burnt and emits a phenolic odor. Grades filled with glass or asbestos are virtually incombustible.

18 Component finishing

Components molded from PF are not susceptible to scuffing and so individual packaging is not normally necessary. Color may be changed by stoving or painting. Hot foiling can be used but is difficult to do well. Components may be electroplated. Flash is commonly removed by linishing or filing: holes or threads can be incorporated by drilling and tapping. Components may be joined by adhesives, riveting or self-tapping screws.

19 Other comments

Phenolic resins are now being used in place of polyester resins to allow the production of large, glass-fiber reinforced moldings by low pressure processes such as hand or spray lay-up. Complex moldings like train cab components may be produced without the need for heat, pressure and expensive forming tools. The big advantage is that phenolics are inherently fire resistant and in a fire do not produce copious quantities of smoke and fumes.

20 Typical components

Glass fiber (GF) filled PF has a high mechanical strength in thin sections when compared to other materials. Because of this it is used for parts with thin sections, such as small bobbins, thin-walled connectors and housings. PFs are also used for electrical items such as coil bobbins, terminal blocks, relay bases and component housings. In the home, PF is used for ashtrays, saucepan handles, knobs and percolator bases. The material is chosen for these applications because it is heat-resistant, durable and cheap. The maximum service temperature depends on, for example, the filler used: for maximum heat resistance, mineral-filled grades should be used. Slate, asbestos and other minerals give heat resistant materials whereas ground mica gives the best electrical resistance. By correct formulation, it is possible to produce components which will withstand $280°C/536°F$: such molding materials are used to produce components for ovens at comparatively low cost. Glass or other fibrous fillers give components with the highest impact strength (3 KJm^{-2} as compared to 1.5 for a GP grade). If the moldings contain organic fillers such as woodflour, then under hot, wet conditions, microbiological growths may appear on the surface.

Phenolic, glass reinforced plastics (PF GRP) are now being used in a variety of interior design applications, particularly in the mass transport sector, for example, for interior surfacing or cladding panels in trains and aircraft. Such materials, based on PF resins and fibrous reinforcement (often glass), are formed by techniques developed for UP resins. Moldings can be glossy but are opaque and dark-colored: they have the tremendous advantage of being inherently flame retardant, low smoke systems which do not require halogen or phosphorus-based additives to give them such properties. They are stiff, hard, have low elongations and possess good creep resistance. Paints are available to enhance the appearance without sacrificing the desirable properties, for example, flame retardancy and low smoke emission.

21 Unit Price

The price is based upon 1 ton/tonne lots and, a General Purpose Phenol Formaldehyde (GPPF) grade will be given a price of 1.00; other materials will be compared to the price for this GPPF grade. The price relative to GPPF can also be found in Table 6.

SUGGESTED PROCESSING CONDITIONS FOR PF.

Process	Setting	Unit	Value
Compression	Preheat temperature	°C	75 to 105
	Tool temperature	°C	165 to 185
	Pressure	MNm^{-2}	>15
	Curing time	s	30 to 60
Transfer	Tool temperature	°C	165 to 185
	Transfer pressure	MNm^{-2}	>50
	Curing time	s	40 to 120
Injection	Cylinder temperature	°C	65 to 90
	Nozzle temperature	°C	85 to 120
	Melt temperature	°C	110 to 140
	Screw temperature	°C	65 to 85
	Tool temperature	°C	165 to 195
	Injection pressure	MNm^{-2}	83 to 248
	Back pressure	MNm^{-2}	<7 (eg 0.7)
	Curing time	s	15 to 80

Process	Setting	Unit	Value
Compression	Preheat temperature	°F	167 to 221
	Tool temperature	°F	329 to 364
	Pressure	psi	>2,200
	Curing time	s	30 to 60
Transfer	Tool temperature	°F	329 to 364
	Transfer pressure	psi	>7,250
	Curing time	s	40 to 120
Injection	Cylinder temperature	°F	149 to 194
	Nozzle temperature	°F	185 to 248
	Melt temperature	°F	230 to 284
	Screw temperature	°F	149 to 185
	Tool temperature	°F	329 to 383
	Injection pressure	psi	12,000 to 36,000
	Curing time	s	15 to 80

SHEET MOLDING COMPOUND.

1 Common name(s)

Sheet molding compound.

2 Abbreviations

SMC; PMC.

3 Alternative names

Polyester molding compound (PMC)

4 Some suppliers

BASF
Bayer
Freeman Chemicals
ERF Plastics
Mia Chemicals
SIR
Scott Bader
Scott Bader

5 Trade names and trade marks

Palatal (resins only)
Leguval (resins only)
Flomat
ERF SMC
Miapol (resins only)
Sirester (resins only)
Crystic (resins only)
Crystic Impreg.

PMC compounds.

Ashland Chemical
Chromos - Ro - Polimeri
Dr. Illing
Haysite Industrial Plastics
Isola Werke
Orkem
Pheonix
Resart
Toyobo
Ubbink Nederland

Aropol and Hectron
Esterform
Illandur
Haysite
Durapol
Norsomix
Keripol
Resatherm
Vyloglas
Uromix

6 Material properties

SMC is made from a resin solution, usually by dissolving an unsaturated polyester (UP) resin in styrene. This resin solution, tailored to tight tolerances, is used to make a polyester molding compound, (PMC) by mixing together the resin solution, fillers, thickening agents (also known as maturation agents), mold release system, inhibiting system, a thermoplastics material and, a catalyst system. The mix is spread onto two, moving, plastic sheets (PE or PA) at the same time as glass (usually rovings) is chopped onto the lower sheet.

The two plastic sheets are then passed through the nip of two rolls which consolidates the molding compound and, makes a sandwich (a sheet) of it between the two plastic sheets. The compacted sheet is then wound on to a roll and set aside to thicken or mature at 30°C or, approximately 84°F. This maturing, or "livering", action is brought about by the maturation agent (for instance, magnesium oxide) and, results in a loss of tack and an increase in strength. As a result, the covering sheets can be stripped off just before molding (keep on as long as possible to prevent contamination and inter-sheet adhesion).

The fillers used are usually calcite, dolomite and aluminum trihydrate (ATH): they lower the cost, and control the rheology, so that fiber separation does not occur on molding. A blend of fillers such as ATH and calcite may also be used so as to obtain controlled rheology and flame retardancy. The addition of a thermoplastics material, reduces shrinkage and improves the surface finish – the use of a thermoplastics material results in materials which are called 'low shrink' (LS) or, 'low profile' (LP) resins. UP resins are polar and, because of this, they have an affinity for metal (mold) surfaces; this adhesive tendency can be counteracted by adding a stearate which melts just below cure temperature – magnesium stearate melts at 130°C or approximately 266°F. Catalysts (initiators) are essential for curing and are usually peroxides: mixed peroxide systems are commonly used in thick sectioned moldings where cracking can be a problem. Glass fibers are widely used, for example, in the size range 12.7mm to 51mm, or 0.5 to 2in (usually 25.4mm or 1in) and at standardized concentrations from 15 to 40%.

Considerable improvements in strength, and other properties, result from high glass fiber concentrations: this is why the glass content may reach 65% - although this may only be a local area of reinforcement. Inhibitors, such as hydroquinone, are necessary to give the required storage life and, to stop premature cure when the mold is being loaded.

7   Forms available

SMC is supplied as sheets in roll form, wrapped around cardboard tubes. Roll width varies from 0.5 to 1m, approximately 19.7 to 39.4in, and actual width is dependant upon the size of the article being molded. The weight of a roll is usually 50kg or, approximately 110lb, but rolls of up to 200kg (440lb) have been used. For such heavy rolls, up to 200kg, specialized unrolling equipment is needed to assist in handling, prior to cutting to size/shape. Sheet thickness is approximately 3mm or 0.118in.

When injection molding SMC, the material can be supplied in sheet form and then rolled into "logs" before being placed into a stuffer unit: this unit ensures that the injection molding cylinder, or barrel, is fed with the molding material.

SMC is usually based upon "E" glass chopped rovings: chopped strand mat (CSM) has been used but rovings are cheaper. Glass content is normally 20 to 25%, however for automotive components, the content is increased to 35% and, in special cases, up to 65%. Isophthalic, and orthophthalic acid, based resins are used and all SMC grades are available with medium, low and minimal shrink characteristics. (see Shrinkage)

8 Materials handling, packaging and storage

Storage life is similar to BMC (thickened DMC) at below 20 to 25°C/68 to 122°F, in wrapped rolls. The SMC may be wrapped in Cellophane, which resists styrene loss better than PE. Storage life is less than 3 months - aim for 1 month at a low, constant temperature. Adopt a very strict fi-fo policy. Dry, or granular PMC, usually has a longer storage life than styrene-based materials; granular PMC is usually styrene free. Such molding compounds are often crosslinked with diallyl phthalate.

9 Color possibilities

SMC is often seen in natural colors (for example, grey and beige) as in many applications (for instance, under the bonnet) there is no demand for bright colors. It can be made virtually any color but only at the manufacturing stage, because of the need for good pigment dispersion without fiber degradation. The final moldings are light stable, even with light colored pigments. The color range is restricted in low profile grades (see Shrinkage) to pastel shades because of patchiness: color penetration in the thermoplastic phase is poor. Use low shrinkage SMC (such as with a PS addition) where bright colors are required. The surface of SMC is often spoilt by grubby marks and porosity: powder mold coating (PMC) appears to offer a way around such problems. This can give hard, scratch resistant coatings (by incorporating glass beads) in a wide range of colors.

10 Molding conditions

Like most PMC compounds, SMC flows readily with relatively low pressures at processing temperatures: SMC cures quickly and, as the cure is achieved by addition polymeriza-

77

tion, volatiles are not formed as a by product. SMC is turned into large area components by compression molding as this process maintains fiber integrity. SMC is easy to use, in simple or complex molds, and can accommodate inserts and coarse screw threads - by the use of suitable fixtures in the tool. However, to simplify molding, inserts are often pushed into molded-in holes during finishing.

SMC compression moldings are usually large components such as exterior car parts. Presses and platens must be well engineered so that wall thicknesses are kept uniform. In addition, displacement transducers may be used to measure platen separation at the corners and, the information fed back to a microprocessor based, feedback system. After the press has closed in the usual way, hydraulically actuated, wedges or rams (located at each corner) may be used to make fine adjustments.

When compression molding it is best initially to load the cavities to be molded with a charge whose area is about half the cavity area - and correspondingly thicker. This is because air is displaced as the material flows and also, because the fill resistance is lower. The hot surface of the tool warms up the molding charge, whose viscosity falls, and this low viscosity material is squeezed out radially thus exposing fresh surfaces to heat: heat transfer is therefore more efficient.

During handling avoid skin contact by, for example, wearing gloves; avoid breathing the fumes and, ensure the area is well ventilated. The material is unwound and cut to size: the backing layers are removed and the charge weight adjusted to some +/– 1%. The material is often plied up and, the shape adjusted to suit a particular mold. For a structural component, this may mean trying to cover the mold completely with the charge (to minimize fiber alignment). This 'tailoring' can only be determined after molding trials. Charge placement is of critical importance but once charge weight, shape and size have been established, keep them constant. Mold temperatures of the order of 170° C/338° F, can cause premature cure, so that resistance to flow and mold filling may be higher than it is at a temperature of, say 150° C or 302° F. It is important that homogenous flow occurs during molding, even when the mold has deep draw areas or sectional changes, as a constant resin to glass ratio is required throughout the molding. For very thick moldings (up to 51mm or 2in) mold temperatures of 120° C/248° F are used. High output, thin walled components (less than 6mm or 0.236in) are often molded at 160° C/320° F.

Pressures required range from 500 to 2000 psi (3.4 to 13.8MNm^{-2}) dependant on compound formulations and on, the complexity of the component. The parting force should be

20% of the rated press tonnage. Closing speeds should be high, for example, 600in/min (15.24m/min); followed by slow speeds of less than 20in/min (0.51m/min) for the last inch (25mm) or so. This is judged from the moment the top rod hits the charge and is necessary to avoid air entrapment. Rapid pressure build up, to full pressure, is essential (less than 5 seconds from rest). Pressure affects surface finish and as, compounds age and harden, higher pressures may be needed to get an acceptable surface finish. Reductions in cure time have been achieved by using peroxide mixtures like t-butyl-perbenzoate (TBPB) with t-butylperoctoate: at 150°C/302°F, the curing times could be 90 seconds. Components used for the automotive industry require an "A" type finish (an excellent finish): conventional molding may not produce such a finish and, moldings may therefore require additional treatment to remove, or fill in, the pin holes or areas of porosity. Repair of the molded surface is usually carried out during the finishing stage. When molding components, possessing thin ribs or deep bosses, particular attention should be given to preventing these areas becoming resin rich and, thus extremely weak due, to the lack of reinforcement. Ensure that the flash land area is kept free of cured material.

There is great interest in injection molding SMC as the products are mechanically stronger than DMC – due to longer residual fiber length and higher glass contents. It may be fed into the injection barrel just like rubber strip – cut back the barrel entry region and the screw flights. Feed the SMC strip (approximately 40mm/1.6in, wide) from a drum, down the side of the modified entry port. (Fiber degradation is minimized by using injection-compression techniques). To prevent resin build-up, use non-tacky SMC strip, round (or radius) all the corners and, well polish all metal surfaces such as the barrel walls. If a stuffer unit is available, SMC may be fed like DMC. Screw machines give best surface finish but plunger machines give reduced fiber degradation. However, most fiber degradation takes place in the feed system. As with DMC, poor tool design results in components with poor surface appearance and with low strength. Vacuum venting of the mold is recommended.

11 Mold considerations

Mold design is specialized for these materials as shrinkage and flow are different from conventional thermosets; this must be taken into account when designing the molds. The moldings may also be incredibly large. Semi-positive compression molding is the most widely used technique as air must be excluded while maintaining material flow. For prototype molds, silicon bronze or beryllium copper have been used: chrome plate to avoid inhibiting the curing reaction and to get the best possible surface finish. Aluminum and Kirksite molds are also used. For large scale production, the molds are of pretoughened (P20) or hardened nickel chrome (up to 60RC) steels.

Cast steels are often selected, in preference to forged billets, for deep cavity molds so as to save on machining costs but, surface finish and mold life are not so good. Flame harden the flash edge and use a vertical flash type mold if possible. The longer the parallel section at the flash line, the more positive will be the tool, and the easier it will be to mold fine detail and to get improved surface finish. The length of the parallels can be 25.4mm/1in and the gap may be 0.1mm/0.004in. Provision is usually made to vent the mold at the pin and, for flash to penetrate to a small recess (approximately, three diameters) down the pin. Use pins of large diameter, pins with mushroom heads or valve ejectors – particularly for low shrink compounds which have low hot strengths. Draft angles do not usually exceed 3 degrees. The angle depends on depth of draw - it may be as little as 1 to 2 degrees for a 203mm/8in depth of draw and 2 to 3 degrees if the depth is less than 203mm/8in. Increase the angle if the mold is textured.

Ensure that mold guidance systems are robust and can withstand, without distortion, the forces generated during mold closing and clamping. Rectangular guides can be aligned accurately and resist shear edge wear. Position guides on the mold centerline to minimize thermal expansion problems. Hard chrome plate the mold to minimize abrasion, to ease the release of the molding from the mold and to impart a good surface finish to the molding. Molds are commonly heated by conduction from the platens as this simplifies mold changing. On large molds, accurate and uniform heating is essential to avoid cure and flow problems. This is best done with hot oil as this can remove exothermic heat and, also give mold temperature uniformity. Such tools are, however, very expensive.

12 Ease of flow

Easy flow materials are those which can be compression molded with pressures as low as 35 bar (3.5MNm^{-2}). Ease of flow can alter during storage as the thickening action (between magnesium oxide and resin) continues. Now, molding compounds are available which are thickened by the addition of solid UP resins, giving more consistent products which are moldable immediately after manufacture and, which use much lower molding pressures – 4 bar (0.4MNm^{-2}). Low viscosity SMCs allow a large reduction in the mold and equipment costs needed for large area moldings.

13 Shrinkage

A GP SMC contains no shrinkage control system, is cheap, and is readily colored. Shrinkage may be 0.4% (0.004in/in or mm/mm) and is caused by the high polymerization shrinkage. Adding a thermoplastic reduces shrinkage and improves surface finish but alters properties such as temperature performance and colorability. If materials like PS or styrene butadiene rubbers are added, it is possible to get low shrinkage SMC where the shrinkage is 0.1 to 0.3% (0.001 to 0.003in/in). Adding thermoplastics such as polyvinyl acetate or polymethyl methacrylate gives low profile SMC: these have the lowest shrinkage of all (0.02 to 0%).These are also known as zero shrink SMC. SMC moldings may undergo virtually no post molding shrinkage and are therefore dimensionally stable: this feature also minimizes component cracking.

14 Resistant to

High temperatures and outdoor exposure. SMC resists water, alcohols, aliphatic hydrocarbons, detergents, lubricants, greases and oils. May resist weak acids, esters, benzene and boiling water. SMC has excellent resistance to arcing (compares favorably with epoxides and MF). The modulus drops as the temperature is raised and glass content has little effect. To improve temperature resistance, use heat-resistant resins and continuous glass fibers. High glass contents (up to 70%) improve stress resistance and, the use of ATH can give flame retardancy to UL V-O.

15 Not resistant to

Ketones and chlorinated hydrocarbons: the resistance to aromatic hydrocarbons, strong acids and alkalis is also not very good and decreases as the temperature increases. In their resistance to moisture, polyester molding compounds (PMC i.e. SMC and DMC) are often worse than some reinforced thermoplastics.

16 Tests for cure or stability

On removal from the mold examine the molding for obvious

defects, such as a lack of rigidity, cracks, surface blisters and/or swelling and, a lack of flatness (warpage). Measure the Barcol Hardness, boil in water for one hour, allow to cool thoroughly and then recheck the hardness for under-cure. Commonly, during production, molding times are reduced until the thickest section swells on cooling then, the cure time is increased by around 10 seconds until the defect disappears.

17 Detection or identification

Moldings can be large (up to $4m^2/43ft^2$) and are usually produced by compression molding: this means no sprue or gate mark. All SMC components feel solid and heavy; because of their thickness and density – $1.8gcm^{-3}/1.04oz\ in^{-3}$ and, which is similar to DMC. SMC contains more glass fiber than DMC but less inorganic filler (other ingredients are similar) and, the glass fibers are much longer. Typical composition resin 100, particulate filler 150, glass fiber (approximately 25.4mm/1in length) 80. The surface can be of a very high gloss but can be marred by black smudges caused through incorrect handling of the starting material: moldings are often painted.

Unpainted SMC is difficult to scratch with a fingernail but can be marred with a knife. Difficult to cut with a knife and when cut, small chips form. These materials are more impact resistant than DMC and much more than GPMC; however cracks may still appear when struck with sharp object. Usually colored throughout the molding but color can be only skin deep as IMC (in-mold coloring) is possible and painting is widely performed. (PMC finishes can be very mar resistant.) When placed in a flame, an SMC molding ignites quite easily, and bright yellow flames occur. Upon removal, the flame extinguishes forming white smoke and, an organic smell of styrene.

18 Component finishing or decoration

Components requiring bright colors are normally produced by painting the molded components after molding. (This is done, in particular, with low profile SMC grades as patchiness occurs with dense or dark colors, e.g. red, blue, etc.). For the car industry, SMC components require degreasing, priming, painting and force drying at elevated temperatures as, the painted components must withstand arduous tests, e.g. 10 days immersion in water at $40°C/104°F$. Steam/hot water is used to remove mold releases and dirt then, the components are force dried. Prime with an epoxy or two-part, PU system. Neutralize static build up with ionized air before spray painting. Deflashing and machining of moldings is similar to that of DMC.

19 Other comments

Moldings may contain slight blemishes, for example, voids, pinholes and porosity, and these may be caused through air entrapment. These blemishes may be filled, sanded and then painted, to give sound moldings with an excellent finish. For filling, use a polyester putty which has the same density as the molding and, which has been formulated to set in <10mins at room temperature. Fill large holes in stages and then, finally sand back to the original contour. Fine porosity may be filled by applying acrylic mist lacquers.

20 Typical components

SMC gives moldings with excellent dimensional stability, high mechanical properties (superior to DMC/BMC), good chemical resistance and electrical insulation. Used for the volume production of high strength, compression moldings such as lorry automotive body parts, electrical housings, chemical trays, furniture, light fittings, etc. Minimal shrink grades can, when using suitable tools, produce "low-profile" moldings, i.e. moldings with center-line average values of less than 0.25 microns/0.001in and, exhibiting no long-term waviness. Such moldings may be painted to automotive standards of finish after a suitable degreasing and priming operation. However, in general, the finish on the molding is only as good as the finish on the tool so great care must be taken with final mold polishing and plating; such moldings are expensive. Used for lorry cabs, car bumpers, car bonnets, etc.. SMC moldings, have good electrical insulation properties with high tracking resistance and a small dielectric loss.

Because of blistering/porosity problems, moldings may require considerable finishing, e.g. up to 3 primer coats (with sanding and finishing in-between) to get the required standard of finish. Coating the mold with a UP resin system, reinforced with glass beads, before molding (by powder mold coating) can give very durable finishes and will extend the use of SMC into sanitary ware, sink tops, etc.; that is, into vitreous enamel replacements.

Due to its high inherent strength properties, stiffness and dimensional stability, SMC is used for the manufacture of parabolic antenna dishes: such components are virtually maintenance free and resist distortion even at wind speeds of up to 200km/h or 124mph.

Considerable improvements in strength, and other properties, result from high glass fiber concentrations, for example, 65%. When such high glass contents are combined with high performance, resin matrices then, structural SMC, or high performance SMC, is the result. Moldings made from such materials are moving into applications requiring mechanical strength, temperature resistance and dimensional stability rivalling, or even surpassing, that of metals. The easy production of complex shapes by compression molding also opens metal markets, for example, a wide variety of fairings, covers and access doors are being used in aircraft. Glass is the most common reinforcement but epoxy, vinyl ester and polyimide SMC with carbon fiber are now available. Typically, structural SMC contains 50 to 70% by weight, 40 to 50% by volume, of randomly orientated, chopped fibers in the resin matrix.

Vinyl ester resins, also known as phenylacrylate resins, are used where, for example, very good chemical resistance is required).

Epoxy/epoxide resins offer good chemical resistance and good retention of properties up to 150 to 180°C/302 to 356°F. Moldings have very low porosity and can be molded in sections as thin as 0.5mm/0.02in and as thick as 152mm/6in.

Phenolic resins are used because of their good flammability characteristics together with low smoke and reduced levels of toxic by-products.

Polyimides are of interest because they give moldings which can be used at temperatures greater than 300°C/572°F.

21 Unit price

The price relative to PF (General Purpose Grade) is as follows:

General Purpose grade	1.68
Low Shrinkage (profile) grade	1.78
Flame retardant – low shrinkage	2.18

The price relative to GPPF can also be found in Table 6.

SUGGESTED PROCESSING CONDITIONS FOR SMC

Process	Setting	Unit	Value
Compression	Preheat temperature	°C	not preheated.
	Tool temperature	°C	120 to 170
	Pressure	MNm^{-2}	3.5 to 14
	Curing time	s	1mm to 30 / 2.5mm to 60
Transfer	Tool temperature	°C	Not usually
	Transfer pressure	MNm^{-2}	transfer
	Curing time	s	molded
Injection	Cylinder temperature	°C	20 to 60
	Nozzle temperature	°C	20 to 40 but may require cooling.
	Melt temperature	°C	20 to 50
	Screw temperature	°C	Keep cool
	Tool temperature	°C	140 to 200
	Injection pressure	MNm^{-2}	27.5 to 80.0
	Back pressure	MNm^{-2}	<0.3
	Curing time	s	20 to 90s 30s for a 2mm section

Process	Setting	Unit	Value
Compression	Preheat temperature	°F	not preheated.
	Tool temperature	°F	248 to 338
	Pressure	psi	500 to 2,000
	Curing time	s	0.039in to 30/0.098in to 60
Transfer	Tool temperature	°F	Not usually
	Transfer pressure	psi	transfer
	Curing time	s	molded
Injection	Cylinder temperature	°F	68 to 140
	Nozzle temperature	°F	68 to 104 but may require cooling.
	Melt temperature	°F	68 to 122
	Screw temperature	°F	Keep cool
	Tool temperature	°F	284 to 392
	Injection pressure	psi	4,000 to 11,500
	Back pressure	psi	<50
	Curing time	s	20 to 90s 30s for a 0.079in section

UREA FORMALDEHYDE

1 Common name(s)

Urea formaldehyde molding material.

2 Abbreviation

UF.

3 Alternative name

Aminoplastic: urea: urea resin molding compounds or materials.

4 Some suppliers

American Cyanamid
BIP
Chemiplastica
Huls (Dynamit Nobel)
Perstorp Ferguson
SA Aicar
SIR
SPREA
Sterling Moulding Materials

5 Trade names or trade marks

Beetle
Beetle, Scarab
Urochem
Pollopas
Urea molding compound
Carbaicar
Siritle
Uroplast
Uroplast.

6 Material properties

An aminoplastic is a plastics material formed by the reaction between materials containing amine ($-NH_2$), or amide groups, and aldehydes. The two, commercially important members of this group are UF and MF. To make a UF material, a low molecular weight resin is first produced by reacting urea and formaldehyde and then, this syrup is combined with fillers, lubricants, hardeners, etc. to give a molding powder.

UFs are thermosetting materials which are available in fine powder or granular form. A wide color range is possible, (because of the lack of resin color). When woodflour is used as the filler brown moldings will result: this is masked by intense coloring so as to give black or brown shades. Such molding powders have a well balanced range of properties and are cheap. They are used in electrical and closure applications. If bleached wood pulp (paper) is used as the filler, then a wide color range is possible as the resultant compound can be translucent: bright, intense colors are possible – including pastel shades and white. Such colors are lightfast. The properties are similar to woodflour filled grades and, the moldings are widely used in electrical fittings for their low cost, wide color range, rigidity and good electrical properties.

The heat resistance of UFs is lower than that of PFs but arc resistance is higher and curing time is faster. The water absorption is also significantly higher than for either PF or MF. In order to obtain improved water resistance the resin may be fortified with melamine: acid, alkali and heat resistance are thus improved. Flame resistance, as measured by UL94, is excellent at V-0: oxygen index of typical moldings is approximately 35%. The bulk density is approximately 700kg/m^3 and the bulk factor is of the order 2.2.

7 Forms supplied

Grades are available for all types of processing – automatic/manual compression molding, transfer and injection molding; such materials are commonly supplied in granular form. Some grades are less gassy than others and so may eliminate the need for mold breathing – such grades are also useful for thick-sectioned moldings. If pelleting is to be performed, then a suitable grade should be ordered.

8 Materials handling, packaging and storage

UF is often supplied in 25kg/55.1lb paper sacks which should be stored in a cool, dry place at a temperature below 20°C/68°F. UF will cure quicker in store than PF. However, if protected against changes in moisture content, the shelf life will be at least six months but adopt a strict Fi-Fo policy. Transfer sealed bags to the molding shop at least 12 hours before use, so as to allow temperature equilibration. Keep containers sealed because, as with other thermosets, UFs are hygroscopic and will therefore absorb water: this will affect material flow, preheating behavior and, component performance. Preheat to lower temperatures than PF – because of a shorter useful life at preheat temperatures. A typical, hot air oven (convection preheating) temperature is up to 60°C/140°F and, a typical high frequency (H.F.) preheating temperatures up to 95°C/203°F.

9 Color possibilities

The materials are supplied in a wide range of colors. White tints (from semi-translucent to fully opaque), pastel shades and intense colors (from white to black) are available. Light fast in both white and pastel shades.

10 Molding conditions

Molding temperatures are slightly lower than those recommended for MF. Too low an injection barrel temperature should however be avoided as this will tend to cause precure – due to the development of large amounts of frictional heat. Such heat is unlikely to be uniformly distributed and so may result in highly stressed moldings. As with other thermosets, a reasonably fast injection speed is suggested so as to get rapid cure and a good surface finish: speeds should be slightly lower than those used for PF. Developments in profiled injection have taken place but acceptance is patchy due to the relatively small number of machines that have this facility.

If mold temperatures are too high, blisters will occur on molded surfaces. If porosity is observed the injection speed is too high or, the mold temperature is too low (under 135°C/275°F). Temperatures must be held accurately – more accurately than for PF. UF can also discolor through overcuring of delicate shades.

To avoid overpacking, set the hold pressure to a low value and adjust until sinking around the gate just stops. This is necessary as aminoplastics can be so overpacked that unacceptable internal stresses result. There are now improved injection grades which are not so prone to porosity and internal stresses. Keep the back pressure low, just sufficient to give shot weight control. Adjust the hold pressure to one third, to one fifth of injection pressure, and bring in at 90 to 95% of stroke so as to minimize stresses. Screw rpm should also be low, about 50 rpm.

When compression molding, allow a cure time of about 20 seconds for a 2mm/0.079in thick molding at a temperature of 150°C/302°F: a 4mm/0.157in thick molding will require about 50 seconds. If the material is being preheated by infrared, keep it well stirred or, keep in very shallow layers: when preheated to high temperatures (60 to 90°C/140 to 194°F) by this method, care must be taken so as to prevent powder "lumping" occurring. With UF, it is very easy to confuse undercure with overcure as both may result in the appearance of blisters. Because of the speed of cure, overcure should always be suspected first.

When processing any plastics material which is based on formaldehyde, care must be taken and the fumes evolved treated as harmful: good ventilation of the work-place is essential.

11 Mold considerations

When injection molding aminoplastics, it is recommended that the cavities be electroplated and that due allowance be made for after-shrinkage (caused by water loss). During the hardening, or setting process, a large amount of volatile matter (such as water) is produced. Adequate mold venting must therefore be provided – through pins or, through wide shallow slots around the mold periphery. Venting to atmosphere is essential; chrome plating and draw polishing of vents is also essential as these allow an easier passage for the escape of gases and, for the removal of excess flash.

The technique known as "breathing" may help: just before the mold fills, injection is momentarily stopped and the mold is slightly parted or "bumped". After final mold closing, injection is completed under second-stage pressure. Vacuum evacuation of the cavities can eliminate porosity and eliminate the need for breathing: it permits fast cavity filling and assists the filling of deep indentations or lugs. It is suggested that all new injection tooling incorporate this feature as, if required, it is very costly to incorporate once the mold has been hardened and plated.

For compression molds, venting of cavities can be ineffective because of vent blocking due to the brittleness of the flash. Quite often "double breathing" alleviates the problem of gas entrapment.

Runner systems are small in diameter (3 to 6mm/0.118 to 0.236in) and, particular attention should be paid to cavity layout so as to enable the use of curved junctions, in preference to T-junctions, for multi-cavity work. A variety of gates can be used ranging from film, fan, edge and submarine. The size of gate is dependent upon shape and complexity of component but, is usually kept small and enlarged if required. The size, for example, for a submarine gate may commence at 0.8 to 1.0mm/0.032 to 0.039in diameter. Because gate wear is not as bad as for PF, woodflour filled UF has been considered as a PF replacement. It is still good practice, however, to carry gates on replaceable pads. Each cavity should be surrounded by a land, which is approximately 9mm/0.354in wide, as this will help avoid flashing in injection cavities; do not locate holes or slots, such as dowel pins or cavity inserts close to the land as this may make flash more difficult to remove. During ejection use an air blast to remove loose flash. Warm sprue and warm runner moldings are also used with these materials in an effort to reduce material usage.

As the materials are forced through the barrel/mold assembly they gain heat and this causes the viscosity to change dramatically; they are worse than PFs in this respect. The clamping force should be greater than 2.5tsi (38.75 MNm^{-2}). Easy flow grades may require 4tsi (62 MNm^{-2}) if flashing is to be avoided. On the clamp side a "mold-halt" facility is useful for insert loading.

If alkyds have to be molded after epoxides or phenolics, good cleaning is required; molding with aminoplastics sometimes cleans the compression mold enough. The bulk factor is usually 2 to 3:1 but can be as high as 4:1 with fine powder. When compression molding UF, the horizontal flash semi-positive type mold is preferred; however, the complexity, shape and split line may warrant other designs.

12 Ease of flow

UF comes in a range of flows or grinds, such as easy, medium and stiff for compression and transfer molding. "Easy" is used for transfer molding, "medium" for general purpose, compression molding and "stiff" for the compression molding of large and deep-draw articles. Special injection molding grades are available.

13 Shrinkage

This may reach 1% but is usually 0.5 to 0.8%. After-shrinkage up to 1% will occur on heating and this should be allowed for, if necessary, at the mold design stage. Test moldings by heating them at 80°C/176°F for 48 hours – about 0.5% after-shrinkage should result. (Dimensional stability of aminoplastics is not very good at elevated temperatures, for example, over 80°C/176°F). Shrinkage may be reduced by using a highly-condensed resin to make the molding powder: to maintain ease of flow a plasticizer, such as glyceryl—tolyl-ether, may then also be added.

14 Resistant to

Solvents and household chemicals. Not affected by fats, oils, esters, ether, petrol, alcohol or acetone; nor by detergents or weak acids such as acetic. In general, the acid resistance is not good. Good resistance to weak alkalis.

15 Not resistant to

Strong acids and alkalis. Not as stain resistant as MF.

16 Tests for cure or stability

As with other thermosets, the degree of cure allowed during molding, depends on the properties required and on the economics of the process. A good balance of electrical and mechanical properties results if the molding has good water resistance. To test for this, saw the molding through its thickest section, immerse in boiling water for 10 minutes, remove and inspect. If the surface appears unaffected or, if it cannot be scratched with a fingernail, it is adequately cured. Accelerated ageing tests are used to decide if a molding will crack in-service, for instance heat to 60° C/140° F for 48 hours and then, examine the molding for cracks.

17 Detection or identification

UF moldings have a hard scratch resistant surface with a high gloss. They are commonly seen as white or pastel shades, with black or brown, also widely used. They are not so resistant to strong acids and alkalis as MFs. As the density of this type of material is about $1.5\text{gcm}^{-3}/0.87\text{oz.in}^{-3}$, the moldings sink rapidly in water and in saturated magnesium chloride solution. Cutting and burning are difficult. On burning, swelling and cracking may occur and the edge of the burnt portion turns white. A pale yellow flame is produced and the flame edge is colored pale blue-green. A strong odor of formaldehyde is produced together with a musty odor (smells like burnt fish). Moldings are easily stained on immersion in hot dye solutions: MF resists such immersions. Because of the residual formaldehyde, and their poor stain resistance, UFs are not often seen as eating or drinking utensils.

18 Component finishing

Flash is removed by filing, linishing, automated deflashing, and so on. Moldings readily accept painting and priming. The appearance of compression moldings may be enhanced by molding-in a resin impregnated foil which contains a legend or scene: this then forms an integral part of the molding. To minimize distortion, components may be placed on jigs directly after molding for up to 15 minutes. It is, however, best to adjust the molding conditions, or the mold design, so that internal stresses and the need for jigs are avoided.

19 Other comments

Molding powder use is only a small outlet for UF resins. They are more widely used as adhesives in, for instance, particle board and plywood.

20 Typical components

UFs are used extensively in the electrical industry for insulating parts in plugs, sockets, switches, connectors and lamps.

In domestic applications, UFs find uses as handles and knobs (furniture and kitchen utensils), buttons, buckles and closures for jars and bottles: this is because of its good resistance to solvents etc. and excellent surface finish. They are widely used for toilet seats, because of the bright, attractive appearance possible.

Formaldehyde is very toxic, and carcinogenic, and so free formaldehyde is the major cause for concern when dealing with the use of aminoplastics which come into contact with food. The toxicity of free melamine and of free urea is low in comparison with that of formaldehyde. Standards exist which specify how free formaldehyde is measured in UF resins.

21 Unit Price

The unit price relative to PF (General Purpose Grade) is as follows:

Compression molding grade (white)	0.99
Compression molding grade (colors)	1.24
Compression jmolding grade (special colors)	1.37
Injection molding grade (white)	1.05
Injection molding grade (colors)	1.29
Injection molding grade (special colors)	1.42

The price relative to GPPF can be also found in Table 6.

SUGGESTED PROCESSING CONDITIONS FOR UF

Process	Setting	Unit	Value
Compression	Preheat temperature	°C	<95
	Tool temperature	°C	135 to 165
	Pressure	MNm^{-2}	>15 eg 30 to 60
	Curing time	s	30 to 180
Transfer	Tool temperature	°C	150 to 170
	Transfer pressure	MNm^{-2}	>50 eg 60 to 120
	Curing time	s	30 to 120
Injection	Cylinder temperature	°C	65 to 95
	Nozzle temperature	°C	85 to 110
	Melt temperature	°C	120 to 140
	Screw temperature	°C	50 to 80
	Tool temperature	°C	145 to 165
	Thin sections <3mm		145 to 155
	Thick sections >3mm		135 to 145
	Injection pressure	MNm^{-2}	100 to 248
	Back pressure	MNm^{-2}	<1 (eg 0.5)
	Curing time	s	15 to 80s

Process	Setting	Unit	Value
Compression	Preheat temperature	°F	<203
	Tool temperature	°F	275 to 329
	Pressure	psi	>2,200
	Curing time	s	30 to 80
Transfer	Tool temperature	°F	302 to 338
	Transfer pressure	psi	>7,250
	Curing time	s	30 to 120
Injection	Cylinder temperature	°F	149 to 203
	Nozzle temperature	°F	184 to 230
	Melt temperature	°F	248 to 284
	Screw temperature	°F	122 to 176
	Tool temperature	°F	293 to 329
	Thin sections <0.118in		284 to 311
	Thick sections >0.118in		275 to 284
	Injection pressure	psi	14,500 to 36,000
	Curing time	s	15 to 80

TABLE 1. SOME NAMES AND ABBREVIATIONS OF PLASTICS AND ELASTOMERS.

Common name	Abbreviation
Acetal (homopolymer and copolymer)	POM-H and POM-K
Acrylate styrene acrylonitrile	ASA or AAS
Acrylate modified styrene acrylonitrile	ASA or AAS
Acrylic acid ester rubber	ACM
Acrylonitrile butadiene rubber or nitrile butadiene rubber	NBR
Acrylonitrile butadiene styrene	ABS
Acrylonitrile styrene/chlorinated polyethylene	ACS
Acrylonitrile methyl methacrylate	AMMA
Acrylonitrile styrene/EPR rubber or, acrylonitrile ethylene propylene styrene	AES
Alpha methyl styrene	AMS
Atactic polypropylene	APP or PP-A
Butadiene rubber or, cis-1,4-polybutadiene rubber or, polybutadiene rubber	BR
Butadiene styrene block copolymer	BDS
Butyl rubber	IIR
Bulk molding compound	BMC
Casein formaldehyde	CF
Cellulose acetate	CA
Cellulose acetate butyrate	CAB
Cellulose acetate propionate	CAP
Cellulose nitrate	CN
Chlorinated polyethylene	CPE or CM
Chlorinated polyvinyl chloride	CPVC or, PVC-C
Chloro-polyethylene or, chlorinated polyethylene.	CM or CPE or, PE-C
Chloroprene rubber or, polychloroprene rubber	CR
Chlorotrifluoroethylene ethylene copolymers	ECTFE
Cis-polyisoprene or, cis-1,4-polyisoprene	IR
Coumarone indene resins	CIR
Diallyl phthalate	DAP
Diallyl isophthalate	DAIP
Dough molding compound	DMC
Elastomeric alloy melt processable rubber	EA-MPR
Elastomeric alloy thermoplastic vulcanizate	EA-TPV
Epichlohydrin rubber	CHR
Epoxy or, epoxide	EP
Epoxy or, epoxide, with glass fiber	EP GF
Ethyl cellulose	EC
Ethylene acryic acid	EAA
Ethylene propylene diene monomer (an EPR terpolymer)	EPDM
Ethylene propylene monomer (an EPR copolymer)	EPM
Ethylene propylene rubber	EPR or EP(D)M
Ethylene tetrafluoroethylene copolymer)	ETFE
Ethylene vinyl acetate	EVA
Ethylene vinyl alcohol	EVOH or EVAL

Expanded polystyrene	EPS or PS-E or XP or PS-X
Fiber reinforced epoxy or epoxide	FRE
Fiber reinforced plastic	FRP
Fluorinated ethylene propylene copolymer	FEP or TFE-HFP
Glass mat reinforced plastic	GMT
Granular polyester molding compound	GPMC
High density polyethylene	HDPE or PE-HD
High impact polystyrene	HIPS or TPS or IPS
Isobutene-isoprene rubber	IIR
Linear low density polyethylene	LLDPE or PE-LLD
Liquid crystal polymer	LCP
Low density polyethylene	LDPE or PE-LD
Medium density polyethylene	MDPE or PE-MD
Melamine formaldehyde	MF
Melamine phenol formaldehyde	MPF
Melt processable rubber	MPR or EA-MPR
Methyl methacrylate, styrene/polybutadiene	MBS
Natural rubber	NR
Nitrile butadiene rubber or,	
acrylonitrile butadiene rubber	NBR
Olefin thermoplastic elastomer	TPO
Oriented polyethylene terephthalate	OPET
Oriented polypropylene	OPP
Oriented polystyrene	OPS
Oriented polyvinyl chloride	OPVC
Perfluoroalkoxy copolymers	PFA
Phenol formaldehyde	PF
Phenylene ether copolymer	PEC
Plasticized polyvinyl chloride	PPVC or PVC-P
Polyacrylonitrile	PAN
Polyamide	PA
Polyamide 6 or nylon 6	PA6
Polyamide 11 or nylon 11	PA11
Polyamide 12 or nylon 12	PA12
Polyamide 66 or nylon 66	PA66
Polyamide 610 or nylon 610	PA610
Polyamide-imides	PAI
Polyaryl amide	PAA 6 or PAMXD6
Polybenzimidazole	PBI
Polybutadiene rubber	BR
Polybutylene	PB
Polybutylene terephthalate	PBT
Polycarbonate	PC
Polychloroprene rubber or,	
chloroprene rubber	CR
Polychlorotrifluorethylene	PCTFE
Polyether block amide	PEBA
Polyether ether ketone	PEEK
Polyether ester	PEEL or COPE or YPBO or TEEE
Polyether imide	PEI
Polyether ketone	PEK

Polyether ketone ketone	PEKK
Polyether sulfone or polysulfone	PES or PSU
Polyethylene	PE
Polyethylene-high density	HDPE or PE-HD
Polyethylene-linear low density	LLDPE or PE-LLD
Polyethylene-low density	LDPE or PE-LD
Polyethylene-medium density	VLDPE or PE-VLD
Polyethylene terephthalate	PET
Polyethylene terephthalate glycol	PETG
Polymethyl methacrylate (acrylic)	PMMA
Polynorbornene (rubber)	PNR
Polyimides	PI
Polyoxymethylene or, acetal or, polyformaldehyde POM	
Polyphenylene ether	PPE
Polyphenylene oxide (modified)	PPO or PPO-M or PPE
Polyphenylene sulfide	PPS
Polyphenylene sulfide sulfone	PPPS
Polyphenylene sulfone	PPSU
Polypropylene	PP or PPR
Polypropylene copolymer	PP-K
Polypropylene homopolymer	PP-H
Polypropylene oxide	PPOX
Polystyrene	PS or GPPS
Polytetrafluorethylene	PTFE
Polyurethane	PUR
Polyvinyl acetate	PVA or PVAC
Polyvinyl alcohol	PVAL
Polyvinyl carbazole	PVCZ
Polyvinyl chloride	PVC
Polyvinyl chloride acetate copolymer	PVCA
Polyvinyl pyrolidone	PVP
Polyvinylidene chloride copolymers	PVDC
Polyvinylidene fluoride	PVDF
Polyvinyl fluoride	PVF
Rubber modified polypropylene	RRPP or, RMPP or, PP/EPDM
Rubber reinforced polypropylene	RRPP or, RMPP or, PP/EPDM
Rubber reinforced polypropylene– with crosslinked rubber	TPO-XL (an EA-TPV or TPV)
Silicone plastics	SI
Styrene acrylonitrile copolymer	SAN
Styrene-butadiene rubber	SBR or, GRS
Styrene butadiene styrene block copolymer	SBS
Sheet molding compound	SMC
Styrene acrylonitrile copolymer	SAN
Styrene butadiene block copolymer	BDS or SBB
Styrene butadiene styrene block copolymer – a thermoplastic elastomer	SBS
Styrene butadiene styrene block copolymer –	

a saturated thermoplastic elastomer	SEBS
Styrene maleic anhydride	SMA
Synthetic polyisoprene	IR
Tetrafluorethylene-ethylene copolymers	ETFE
Thick molding compound.	TMC
Thermoplastic elastomer (or rubber)	TPE or TPR
Thermoplastic elastomer - amide based	TPE-A
Thermoplastic elastomer - olefin based	TPE-O
Thermoplastic elastomer - urethane based	TPE-U
Thermoplastic ether ester (COPE or PEEL)	TP-EE
Thermoplastic ethylene propylene rubber – a TPE TP-EPDM	
Thermoplastic ethylene vinyl acetate elastomer – a TPE	TP-EVA
Thermoplastic natural rubber - a TPE	TP-NR
Thermoplastic nitrile butadiene rubber	TP-NBR
Thermoplastic polyolefin	TPO
Thermoplastic polyolefin- with crosslinked rubber TPO-XL	
Thermoplastic polyurethane	TPU
Thermoplastic rubber (or elastomer)	TPR or TPE
Thermoplastic vulcanizate (a TPE)	TPV or EA-TPV
Unplasticized polyvinyl chloride	UPVC or PVC-U
Unsaturated polyester	UP
Unsaturated polyester, with glass fiber	UP GF (PMC, DMC or SMC)
Urea formaldehyde	UF
Very low density polyethylene	VLDPE or PE-VLD
Vinyl chloride ethylene	VCE
Vinyl chloride ethylene vinyl acetae	VCEVA
Vinyl chloride vinyl acetae	VCVA
Vinyl ester resins	VE
Vinylidene chloride, acrylonitrile copolymers	VCA

TABLE 2A. LETTERS USED TO MODIFY ABBREVIATIONS FOR PLASTICS (ISO AND ASTM).

Letter	Meaning or significance
C	Chlorinated
D	Density
E	Expandable or expanded
F	Flexible or fluid/liquid state
H	High
I	Impact
L	Linear or low
M	Medium or molecular
N	Normal or novolak
P	Plasticized or plasticised
R	Raised or resol
T	Thermoplastic
U	Ultra or, unplasticized or unplasticised
V	Very
W	Weight
X	Crosslinked or crosslinkable

TABLE 2B. COMMONLY-USED LETTERS USED TO MODIFY ABBREVIATIONS FOR PLASTICS (I.E. IN ADDITION TO TABLE 2A).

Letter	Meaning or significance
A	Atactic or, amorphous
B	Block copolymer
C	Crystalline
E	Emulsion (polymer)
F	Fiber/fibre
G	Glass
H	Homopolymer
K	Copolymer
M	Mass or bulk (polymer) or, mat
O	Oriented
R	Random copolymer or, reinforced
S	Suspension (polymerization)
V	Vulcanized or crosslinked
EP	Engineering thermoplastic or, engineering thermoplastics material
FR	Flame retardant
GF	Glass fiber
GMT	Glass mat (reinforced) thermoplastics (material)
GP	General purpose
HI	High impact
PMC	Polyester molding compound
TP	Thermoplastic
XL	Crosslinked or, cured or, vulcanized

TABLE 2C. SYMBOLS USED FOR FILLERS AND/OR REINFORCING MATERIALS.

Letter	Meaning or significance
A	Asbestos
B	Boron or, beads or, spheres or, balls
C	Carbon or, chips, or, cuttings
D	Powder
F	Fiber/fibre
G	Glass or, ground
H	Whisker
K	Chalk or, knitted fabric
L	Cellulose or, layer
M	Mineral or, metal
N	Non-woven fabric (usually thin)
P	Mica or, paper
Q	Silicon
R	Aramid or, roving
S	Synthetic, organic or, scale, flake
T	Talcum or, cord
V	Veneer
W	Wood
Y	Yarn
Z	Others.

Please note that if a letter is not being used in these tables then, it does not mean that it is not being used in another branch of the plastics industry.

TABLE 3. SOME ABBREVIATIONS AND TRADE NAMES OF PLASTICS.

Abbreviation	Common name	Common trade names or, trade marks
ABS	Acrylonitrile butadiene styrene	Cycolac; Lustran
AMS	Alpha methyl styrene	Elite HH
ASA	Acrylonitile styrene acrylonitrile (AAS)	Luran S
BDS	Butadiene styrene block copolymer	K resin; Styrolux
BMC	Bulk moulding compound	Freemix; Norsomi
CA	Cellulose acetate	Cellidor; Tenite
CAB	Cellulose acetate butyrate	Cellidor; Tenite
CAP	Cellulose acetate propionate	Cellidor; Tenite
CF	Casein formaldehyde	Erinoid; Lactoi
CN	Cellulose nitrate	Celluloid; Xylonite
COPE	Polyether ester elastomer	Arnitel; Hytrel
CP	Cellulose propionate (CAP)	Cellidor; Tenite
CPE	Chlorinated polyethylene (PE-C)	Bayer CM; Tyrin CM
CPVC	Chlorinated polyvinyl chloride	(PVC-C)
DAP	Diallylphthalate	
DAIP	Diallyl isophthalate	
DMC	Dough moulding compound	Beetle DMC; ERF DMC
EA-MPR	Elastomer alloy melt processable rubber	Alcryn
EA-TPV	Elastomer alloy thermoplastic vulcanizate	Lomod; Santoprene
EP	Epoxide or epoxy (cured)	Araldite
ETFE	Tetrafluorethylene-ethylene copolymers	Tefzel
EVA	Ethylene vinyl acetate copolymer (EVAC)	Evatane
EVAL	Ethylene vinyl alcohol copolymer	Clarene; Eval
EVOH	Ethylene vinyl alcohol copolymer	Clarene; Eval
FEP	Fluorinated ethylene propylene (TFE-HFP)	Teflon FEP
GPMC	Granular polyester moulding compound	Freeflo; Impel
HDPE	High density polyethylene (PE-HD)	Lupolen HD; Rigidex
HIPS	High impact polystyrene (TPS or IPS)	Lustrex; Polystyrol
LCP	Liquid crystal polymer	Vectra; Xydar.
LDPE	Low density polyethylene (PE-LD)	Alathon; Hostalen LD
MDPE	Medium density polyethylene (PE-MD)	Fortiflex
MF	Melamine formaldehyde	Melmex; Melopas
MPR	Melt processable rubber	Alcryn
PA	Polyamide or nylon.	
PA 6	Polyamide 6 or nylon 6	Akulon K; Ultramid
PA 11	Polyamide 11 or nylon 11	Rilsan B
PA 12	Polyamide 12 or nylon 12	Rilsan A; Grilamid
PA 46	Polyamide 46 or nylon 46	Stanyl
PA 66	Polyamide 66 or nylon 66	Maranyl; Zytel
PA 610	Polyamide 610 or nylon 610	Brulon; Perlon
PAA 6	Polyaryl amide or, poly-m-xylene-adipamide (PA MXD6)	Ixef

Abbrev- iation	Common name	Common trade names or, trade marks
PAN	Polyacrylonitrile	Acrilan; Barex; Orlon
PBI	Polybenzimidazole	Celazole
PBT	Polybutylene terephthalate	Pocan; Valox
PC	Polycarbonate	Lexan; Makrolon
PCTFE	Polychlorotrifluorethylene	Hostaflon C2; Kel-F
PE	Polyethylene	Alathon; Lupolen
PEBA	Polyether block amide	Pebax
PEEK	Polyether ether ketone	Victrex PEEK;
PEEL	Polyether ester (YPBO)	Arnitel; Hytrel
PE-HD	Polyethylene-high density	Lupolen HD; Rigidex HDPE
PEI	Polyether imide	Ultem.
PEK	Polyether ketone	Hostatec
PEKK	Polyether ketone ketone	(from DuPont)
PE-LD	Polyethylene- low density	Alathon; Lupolen
PE-MD	Polyethylene- medium density	Fortiflex
PE-VLD	Polyethylene- very low density	Norsoflex
PET	Polyethylene terephthalate	Arnite A; Techster E
PES	Polyether sulphone	Victrex.
PMC	Polyester molding compound	Aropol; Norsomix
PF	Phenol formaldehyde	Bakelite; Sternite
PI	Polyimide	Vespel
PMMA	Polymethyl methacrylate (acrylic)	Diakon; Plexiglas
PMMA-T	Toughened acrylic	
POM	Polyoxymethylene or, acetal or, polyformaldehyde	Delrin; Hostaform
POM-H	Acetal homopolymer	Delrin and Delrin II
POM-K	Acetal copolymer	Hostaform; Ultraform
PP	Polypropylene	Profax; Propathene
PPE	Polyphenylene ether (see PPO)	
PPO	Polyphenylene oxide – usually modified polyphenylene oxide (PPO-M)	Luranyl; Noryl
PPS	Polyphenylene sulphide	Fortron; Ryton
PPPS	Polyphenylene sulphide sulphone	Ryton S
PPVC	Plasticised polyvinyl chloride (PVC-P)	Solvic; Vinnol
PS	Polystyrene (GPPS)	Lustrex; Polystyrol
PSU	Polysulphone	Udel
PTFE	Polytetrafluorethylene	Fluon; Teflon
PVC	Polyvinyl chloride	Corvic; Geon
PVDC	Polyvinylidene chloride copolymers	Saran
PVDF	Polyvinylidene fluoride	Dyflor; Kynar; Solef
PVF	Polyvinyl fluoride	Tedlar
SAN	Styrene acrylonitrile copolymer	Lustran SAN; Tyril

Abbrev-iation	Common name	Common trade names or, trade marks
SMC	Sheet moulding compound	ERF SMC; Flomat
UPVC	Unplasticised polyvinyl chloride (PVC-U)	Corvic; Geon
RMPP	Rubber modified polypropylene (PP/EPDM)	Uniroyal TPR; Keltan
RRPP	Rubber reinforced polypropylene (PP/EPDM)	Uniroyal TPR; Keltan
SBS	Styrene butadiene styrene block copolymer	Cariflex TR; Solprene
SEBS	Styrene butadiene styrene block copolymer (saturated)	Cariflex
TPE	Thermoplastic elastomer (rubber)	
TP-EE	Thermoplastic elastomer – ether ester	Arnitel; Hytrel
TP-EPDM	Thermoplastic elastomer – based on EPDM	
TP-EVA	Thermoplastic elastomer – based on EVA	
TP-NBR	Thermoplastic elastomer - based on NBR	
TPO	Thermoplastic polyolefin	Propathene OTE; Vistaflex
TPO-XL	Thermoplastic polyolefin rubber – crosslinked (rubber)	Levaflex; Santoprene
TPR	Thermoplastic rubber (elastomer)	
TPU	Thermoplastic polyurethane	Elastollan; Estane
TPV	Thermoplastic elastomer or rubber – crosslinked (rubber)	Lomod; Santoprene
UF	Urea formaldehyde	Beetle; Scarab
VE	Vinyl ester resins	
VLDPE	Very low density polyethylene	Norsoflex.

TABLE 4. SOME TRADE NAMES, ABBREVIATIONS AND SUPPLIERS OF POLYMERS AND POLYMER COMPOUNDS.

Trade name/ trade marks	Abbreviation	Supplier
A-C	Low mol wt. PE	Allied Signal Imc.
A-Clyn	Low mol wt. ionomers	Allied Signal Imc.
Acrylite	PMMA	Canada Colors & Chemicals
Adpro	PP	Advanced Global Polymers
A-Fax	PP-A	Himont
Akulon K	PA6	Akzo Engineering Plastics
Akulon M	PA6	Akzo Engineering Plastics
Alathon	PE-HD	DuPont
Alcoryl	ABS	Rhone Poulenc
Alcryn	EA-TPV or TPE	DuPont
Algoflon	PTFE	Enimont
Altulite	PMMA	Altulor
Amoco HDPE	PE-HD	Amoco
Apec	Polyester PC	Bayer
Appryl	PP	Appryl
Apscom	Thermoplastics compounds	Akzo Engineering Plastics
Araldite	EMC or EP	Ciba Geigy
Ardel	Polyarylate	Amoco
Arnite A	PET	Akzo Engineering Plastics
Arnitel	PEEL or COPE	Akzo Engineering Plastics
Arnite	PBT	Akzo Engineering Plastics
Aropol	PMC	Ashland Chemical Co.
Arylon T	ABS/PSU	USS Chemicals
Ashlene	PA 66	Ashley Polymers Inc
Azdel	GMT/PP	GEP (General Electric Plastics)
Azloy	GMT/PC/PBT	GEP (General Electric Plastics)
Azmet	GMT/PBT	GEP (General Electric Plastics)
Bakelite	PF	Bakelite/Sterling Moulding Materials
Bakelite Polyester Alkyd	GPMC	Sterling Moulding Materials

Trade name/ trade marks	Abbreviation	Supplier
Bapolan	PS	BASF/Bamberger
Bapolan	HIPS	BASF
Bapolene	HDPE & PP	Bamberger
Barex	PAN	Sohio
Barex	PAN	Standard Oil
Bayblend	ABS/PC	Bayer
Bayer CM	CPE	Bayer
Baycomp	Fiber filled thermoplastics	Baycomp
Baygal	Encapsulating resins	Bayer
Baymidur	Encapsulating resins	Bayer
Baypren	CR	Bayer
Beetle	UF	BIPChemicals/ Cyanamid
Beetle DMC	DMC	BIP Chemicals
Beetle nylon 6	PA 6	BIP Chemicals
Beetle nylon 66	PA 66	BIP Chemicals
Beetle PET	PET	BIP Chemicals
Bergamid A	PA66	Bergmann
Bergamid B	PA6	Bergmann
Bergacell	CA	Bergmann
Bexloy	Engineering thermoplastics for cars	DuPont
Buna AP	EPM/EPDM	Bunawerke Huls
Buna CB	BR	Bayer
Buna EM	SBR	Bunawerke Huls
Cabelec	PP (conductive)	Cabot Plastics
Cadon	SMA	Monsanto
Cadon 300	Impact modified SMA (used with PVC)	Monsanto
Calibre	PC	Dow
Caprolan	TPU	Elastogran/BASF
Capron	PA6	Allied Corp
Carbaicar	UF	S A Aicar
Carbopol	Acrylic acid polymer	BF Goodrich
Carboset	Acrylic polymer	BF Goodrich
Carinex	HIPS	Shell
Cariflex	TPE (SBS)	Shell
Celanese Nylon	PA 66	Hoechst/Hoechst Celanese
Celanex	PBT	Hoechst/Hoechst Celanese
Celazole	PBI	Hoechst/Hoechst Celanese
Cellidor	CA & CAP	Albis
Celsir	UF resins	SIR
Celstran	Long fiber filled thermoplastic	Hoechst/Hoechst Celanese
Cellobond BP	PF	BP Chemicals

Trade name/ trade marks	Abbreviation	Supplier
Centrex	ASA or AAS	Monsanto
Clarene	EVOH or EVAL	Solvay
Corton	PP mineral filled	Poly Pacific Pty
Corvic	PVC	EVC
Crastine	PBT	Ciba Geigy
Craston PPS	PPS	Ciba Geigy
Crystic	UP	Scott Bader
Crystic Impreg	GMC	Scott Bader
Cycolac	ABS	Monsanto
Cymel	MF	Cyanamid
DAP 5000	DAP	Synres Amoco
DAIP 6000	DAIP	Synres Amoco
Daplen	LDPE: HDPE: PP	Chemie Linz
Delrin	POM-H	DuPont
Delrin II	POM-H	DuPont
DER	EP	Chemroy Canada
Desmopan	TPU	Bayer
Dexel and Dexel S	CA	Courtaulds Speciality Plastics
Dexflex	TPO	Dexter Plastics
Diakon	PMMA	ICI
Dialac	ASA	Mitsubishi
Dion	UP	Fiberglass
Dowlex	PE-LD	Dow Chemicals
Dowlex	PE-LLD	Dow Chemicals
DSM Nyrim	Reaction injection moulding (RIM) PA	DSM RIM Nylon
Duracryn	TPE/TPV	DuPont
Dural	reinforced UPVC	Dexter Plasics
Duralex	PVC/PU/NBR alloy	Dexter Plasics
Duralon	PA 11	Thermoclad
Durapol	PMC	Isola Werke
Durez	PF	Occidental/ Canadian Oxy Chemicals
Durez	DAP	Occidental
Durolon	PC	Atochem
Dutral	PS	Enimont
Dutral FLX	TPE/TPV	Enimont
Dyflor	PVDF	Dynamit Nobel
Dyflor	PVDF	Kay Fries
Dynaform	RRPP	Dynamit Nobel
Dynaset	PF	Reichold
Ecdel	COPE or PEEL	Eastman Chemicals
Eccomold	EP	Emmerson and Cuming
Ecolyte II & IV	PE	Ecoplastics/ Eco Chemicals

Trade name/ trade marks	Abbreviation	Supplier
Ecolyte	Photodegradable TP	Ecoplastics/Eco Chemicals
Ecolyte S	PS	Ecoplastics/Eco Chemicals
Edister	ABS	Enimont
Edister	HIPS	Enimont
Edister	PS	Enimont
Elastar	TPE (based on NBR-XL/PVC)	
Elastollan	TPU	Elastogran/BASF
Elastopreg	GF/GMT laminates	BASF/Elastogran
Elemid	PPO-M/PA	Borg Warner
Elite HH	AMS	Monsanto
Elvanol	EVAL	DuPont
Elvax	EVA copolymer	DSM
Empee PP	PP	Monmouth Plastics
Envex	Reinforced PI	Rogers Corp
Epoester	EP resins	SIR
Epolan	ABS	Industrial Resistol
Epolene	Maleated PP	Eastman Chemicals
Epon	EP	Shell
Eponac	EP	SPREA
Eposir	EP resins	SIR/ Kingsley & Keith
ERF DMC	DMC	ERF
ERF SMC	SMC	ER
Ertalan	cast PA	Erta
Escorene	PE-LLD	Exxon
Estane	TPU	Goodrich
Estamid	PEBA/TPE	Dow
Esterform	PMC	Chromos Ro-Polimeri
ETA	TPE	Republic Plastics
Europrene	TPE (SBS)	Ani
Eval	EVOH or EVAL	Kuraray/EVAL Co.
Exxelor	Plastics modifiers	Exxon
Fenochem	PF	Chemiplastica Spa
Fenoform	PF	Chromos Ro-Polimeri
Fibiter	PBT	Enimont
Flomat	SMC	Freeman Chemicals
Foraflon	PVDF	Atochem
Fortiflex	PE-MD	Soltex Polymer Corp.
Fortiflex	PE-HD	Soltex Polymer Corp.
Fortilene	PP	Soltex Polymer Corp.
Fortron	PPS	Hoechst/Hoechst Celanese
Freeflow	GPMC	Freeman Chemicals
Flowmat	SMC	Freeman Chemicals

Trade name/ trade marks	Abbreviation	Supplier
Gaflex	COPE or PEEL	Hoechst/Hoechst Celanese
Gedex	PS	Orkem
Gedex	HIPS/TPS	Orkem
Geloy	ASA	General Electric Plastics/GEP
Genal	PF	General Electric Plastics/GEP
Geolast	TPV or NBR-TPV	Monsanto
Geon	PVC	BF Goodrich
Glilax	PEBA/TPE	Dianippon
Grilamid	PA12	EMS-Grilon (EMS-Chemie)/Emser Industries
Grilon	PA 6	EMS-Grilon (EMS-Chemie)/Emser Industries
Grilon T	PA 66	EMS-Grilon (EMS-Chemie)/Emser Industries
Grivory	Amorphous EP	Emser Ind.
Halar	ECTFE	Ausimont
Halon	PTFE	Ausimont
Haysite	PMC	Haysite Reinforced Plastics
Hercules HPR	PE-HD of high molecular weight	Hercules
Hetron	PMC	Ashland Chemical Corp.
Hostaform	POM-K	Hoechst/Hoechst Celanese
Hostadur	PBT	Hoechst/Hoechst Celanese
Hostaflon FEP	FEP	Hoechst/Hoechst Celanese
Hostalen	PE-HD	Hoechst/Hoechst Celanese
Hostalen GUR	PE-UHMW	Hoechst/Hoechst Celanese
Hostalen PP	PP	Hoechst/Hoechst Celanese
Hostalen PP	reinforced PP	Hoechst/Hoechst Celanese
Hostapren	PE-C	Hoechst/Hoechst Celanese
Hostalit	PVC	Hoechst/Hoechst Celanese
Hostalit Z	PVC-HI	Hoechst/Hoechst

Trade name/ trade marks	Abbreviation	Supplier
Hostatec	PEK	Hoechst/Hoechst Celanese
Huntsman Polystyrene		
	PS	Huntsman
Hytrel	PEEL or COPE	DuPont
Idemitsu Polycarbonate		
	PC	Idemitsu Petro Chem.
Illandur	PMC	Dr. Illing
Impel	GPMC	Scott Bader
Impet	PET	Hoechst/Hoechst Celanese
Indopol	Polybutenes	Amoco
Innovex	PE-LLD	BP
Isomin	MF	Perstorp
Ixef	PAA 6	Laporte/Solvay
Iztavil	PVC	Polimeros De Mexico
Kadel	Polyketone	Amoco
Kamax	Acrylic imides	Rohm and Haas
Kane Ace	CPVC	Kaneka/ E W Seward Ltd.
Kane Ace B	MBS	Kaneka/ E W Seward Ltd.
Kane Ace PA	Acrylic proc. aid	Kaneka/ E W Seward Ltd.
Kane Ace XEL	Cross linked PVC	Kaneka/ E W Seward Ltd.
Kelburon	RRPP/TPE	DSM (Dutch State Mines)
Kelrinal	CM rubber	DSM (Dutch State Mines)
Keltan TP	RRPP/TPE	DSM (Dutch State Mines)
Keltan	EP(D)M	DSM (Dutch State Mines)
Kelprox	TPE	DSM (Dutch State Mines)
Keltaflex	TPE/TPV	DSM (Dutch State Mines)
Kematal	POM-C	Hoechst/Hoechst Celanese
Kerimid	PI	Rhone Poulenc
Keripol	PMC	Pheonix
K F	PVDF	Kureha
Kinel	PI	Rhone Poulenc
Kodapek PET	PET	Eastman Chemicals
Kodar PETG	PET-K or, PET-C	Eastman Chemicals

Trade name/ trade marks	Abbreviation	Supplier
Kostil	SAN	Enimont
Kraton TR	TPE (SBS)	Shell
K resin	BDS	Phillips
Kynar	PVDF	Penwalt
Lacqrene	PS	Atochem
Lacqrene	HIPS	Atochem
Lacqtene HD	PE-HD	Atochem
Lacqtene HX	PE-LLD	Atochem
Ladene	PE-LLD	Sabic
Legupren	UP	Bayer
Leguval	UP resins	Bayer
Lekutherm	EP	Bayer
Levaflex	TPO-XL	Bayer
Levapren	EVA/PVC	Bayer
Lexan	PC	GEP (General Electric Plastics)
Linpac Polystyrene	PS	Linpac
Lomod	COPE or PEEL	GEP
Lotader	Co & terpolymers	Orkem (was CdF Chemie)
Lotrene	PE-LD	Orkem
Lotrex	PE-LLD	Orkem
Lucalor	CPVC	Orgavyl
Lucovyl	PVC	Atochem
Lupolen	PE-LD	BASF
Lupolen HD	PE-HD	BASF
Luranyl	PPO-M or PPE/HIPS	BASF
Luran	SAN	BASF
Luran S	ASA	BASF
Lustran	ABS	Monsanto
Lustran	SAN	Monsanto
Lustran Elite HH	AMS	Monsanto
Lustran Ultra ABS	ABS (high gloss)	Monsanto
Lustrex	PS	Monsanto
Lustrex	HIPS	Monsanto
Luran	SAN	BASF
Luran S	ASA or AAS	BASF
Magnum	ABS	Dow
Makroblend	PC/PBT	Bayer
Makrolon	PC	Bayer
Maranyl	PA 66	ICI/LNP
Marlex	PE-HD	Phillips
Melaicar	MF	S A Aicar
Melamine moulding compounds	MF	Perstop Ferguson
Melmex	MF	BIP
Meloplas	MF	Ciba Geigy
Melsprea	MF	SPREA

Trade name/ trade marks	Abbreviation	Supplier
Menzolit	PMC	Menzolit Werke
Merlin	PC	Mobay
Merlon	PC	Bayer
Metton	A liquid moulding resin	Shell/Hercules
Miapol	UP resin	Mia Chemical
Mindel	PSU (filled)	Amoco Chemicals
Minlon	PA 66 (filled)	DuPont
Moldsite	PF	SPREA
Moplen RO	PE-HD	Enimont
Moplen SP	RRPP	Enimont
Morthane	TPU	Morton Thiokol Inc.
Neoflon	FEP	Daikin
Neonite	EMC or EP long glass fiber filled	Ciba Geigy
Neste HDPE	PE-HD	Neste
Neste LD	PE-LD	Neste
Neste LPLD	PE-LLD	Neste
Neste PS	HIPS	Neste
Nestorite	PF	Perstorp
Nike	CN	Punda Inc.
Norchem	PE-HD, PE-LLD & PP	Enron/Delong Prochem
Norlin	PE-LLD	Northern Petrochemicals
Norsomix	DMC/PMC	Orkem (CdF Chimie)
Norsoflex	VLDPE	Orkem (CdF Chimie)
Nortuff	PP	Norchem Inc
Norvinyl	PVC	Norsk Hydro
Noryl	PPO-M	GEP(General Electric Plastics)
Noryl GTX	PPO-M/PA	GEP(General Electric Plastics)
Novamid	a PA	Mitsuibishi
Novablend	PVC	Novatec
Novapol HD	PE-HD	Novacor Chemicals
Novapol LD	PE-LD	Novacor Chemicals
Novex	PE-LD	BP Chemicals
Novodur	ABS	Bayer
Nuloy	PA6	Terlon Polimeros
Nydur	PA6 & PA66	Mobay
Nylafil	Filled PA	Wilson Fibrefil
Nylon	PA	e.g. DuPont
Nypel	PA6	Allied Signal Inc.
Orgalloy R	PA6/PP or PA66/PP	Atochem
Orgamide	PA 6	Atochem
Orgater	PBT	Atochem
Oroglas	PMMA	Rohm and Haas

Trade name/ trade marks	Abbreviation	Supplier
Palatal	UP	BASF
Paraloid	Impact modifiers	Rohm and Haas
Parr	DAP	US Prolam Inc.
Paxon	PE-HD	Allied
Pax-Purge	Purge compounds	Canada Color & Chemicals
Pebax	PEBA a TPE	Atochem
Pemex	LDPE	Petroleos Mexicanos
Pekema	PVC	Punda Inc
Pekevic	PVC	Neste Oy
Perbunan N	NBR	Bayer
Petlon	PET	Bayer/Mobay
Petra	PET	Allied Signal Inc.
Petrothene	PE-HD	USI/Quantum
Petrothene	PE-LLD	USI/Quantum
Pevikon	PVC	Norsk Hydro
Pibiflex	PEEL or COPE	Dutral/Montedison
Pibiter	PBT	Dutral/Montedison
Plasticlean	Purging compound	W S Wood Assoc.
Plaskon	EP	Plaskon Molding Div.
Plaskon DAP	DAP	Plaskon Molding Div.
Plenco	PF, MPF & UP	Plastics Engineering Co.
Plexiglas	PMMA	Rohm and Haas
Pocan	PBT	Bayer/Mobay
Polloplas	UF	Dynamit Nobel
Polyclear	PET	Hoechst/Hoechst Celanese
Polychem	DAP	Budd Co.
Polycol	PVC	Atochem
Poly DAP	DAP	US Prolam Inc.
Polyfort	PP filled	Schulman Inc
Polykemi	PC	Atochem
Polyloy	PA 6	Illing
Polyloy	PA 66	Illing
Polymer E	LDPE	Asia Polymer Corp
Polyset	EP	Morton Chemical Div.
Polystyrol	PS	BASF
Polystyrol	PS	Norsk Hydro
Polystyrol	HIPS	BASF
Polystyrol	HIPS	Norsk Hydro
Polystyron	PS	Svenska
Polyvest	Polybutadiene	Huls
Prevex	PPO-M or PPE	Borg Warner
Primax	PE-UHMW	Air Products and Chemicals
Procom	PP compounds	ICI
Profax	PP	Himont/Hercules
Progilite	PF	Rhone Poulenc
Proloy	EP	GEP(General Electric Plastics)

Trade name/ trade marks	Abbreviation	Supplier
Propathene OTE	RRPP	ICI
Pulse	ABS/PC	Dow
Quimcel	CN	Punda Inc.
Radel	PSU	Amoco
Radlite	GMT	Azdel Europe
Ravikral	ABS	Enichem
Resarit	PMMA	Resart
Resartherm	PMC	Resart
Resilon	PVC	Canadian General-Tower
Resin 18	AMS	Amoco
Rexene	PP	El Paso
Rigidex HDPE	PE-HD	BP
Rilsan A	PA 12	Atochem
Rilsan	PA 11	Atochem
Riteflex	PEEL or COPE	Hoechst/Hoechst Celanese
Ronfalin	ABS	DSM (Dutch State Mines)
Ronfaloy	ABS alloy	DSM (Dutch State Mines)
Ronfaloy E	ABS/EP(D)M	DSM (Dutch State Mines)
Ronfaloy V	ABS/PVC	DSM (Dutch State Mines)
Rovel	Styrene based material	Dow
Royalene	EPDM	Uniroyal
Rutaform	MF	Sterling Moulding Materials
Rutaform Polyester	GPMC	Sterling Moulding Materials
Rutamid 6	PA6	Bakelite
Rutamid 66	PA66	Bakelite
Rynite	PET	DuPont
Ryton	PPS	Phillips Chemicals
Ryton S	PPPS	Phillips Chemicals
Saran	PVDC or PVdC	Dow
Santoprene	TPO-XL an EA-TPV	Monsanto
Scarab	UF	BIP
Scarnol	EVOH or EVAL	Nippon Gobsei
Sclair	PE-LLD	DuPont
Sclairlink	Cross linkable PE	DuPont
Selar	amorphous PA	DuPont
Sinvet	PC	Enimont
Sirester	UP resins	SIR (Societa Italiana Resine)

Trade name/ trade marks	Abbreviation	Supplier
Sirfen	PF resins	SIR (Societa Italiana Resine)
Sirfen X	PF	SIR (Societa Italiana Resine)
Siritle	UF	SIR (Societa Italiana Resine)
SMA Resins	SMA	Sartomer
Siritle	UF	SIR (Societa Italiana Resine)
Solef	PVDF	Laporte
Solef	PVDF	Solvay
Sol T	SBS & SIS	Enimont
Stamylan HD	PE-HD	DSM (Dutch State Mines)
Stamylan LD	PE-LD	DSM (Dutch State Mines)
Stamylan P	PP	DSM (Dutch State Mines)
Stamylex	PE-LLD	DSM (Dutch State Mines)
Stamyroid	modified amorphous PP	DSM (Dutch State Mines)
Stanyl	PA 46	DSM (Dutch State Mines)
Stapron S	rubber modified SMA	DSM (Dutch State Mines)
Statoil	PE-HD	Statoil
Statoil polyethylene LDPE	PE-LD	Statoil
Sternite	PF	Sterling Moulding Materials
Sternite	PS	Sterling Moulding Materials
Sternite	HIPS	Sterling Moulding Materials
Styrolux	BDS	BASF
Styron	PS	Dow
Styron	HIPS	Dow
Supec	PPS	GEP (General Electric Plastics)
Supraplas	EP	Sud West Chemie
Suramin	UF and MF resins	SIR (Societa Italiana Resine)
Surlyn	Ionomer resins	DuPont
Tancin	PP	Washington Penn Plastics
Technyl B	PA 66	Rhone Poulenc
Techster E	PET	Rhone Poulenc

Trade name/ trade marks	Abbreviation	Supplier
Techster T	PBT	Rhone Poulenc
Tecolit	PF	Toshiba Chemical Products
Tedur	PPS	Bayer
Teflon FEP	FEP	DuPont
Teflex	FEP	Nitechim
Tenite	CA	Eastman Chemical
Tenite	CAB	Eastman Chemical
Tenite	CP or CAP	Eastman Chemical
Tenite polyethylene	PE-LD	Eastman Chemicals
Terblend S	ASA/PC	BASF
Terluran	ABS	BASF
Therban	NBR or HNBR	Bayer
Thermocomp	PES/PSU	LNP Plastics
Thermocomp	Filled PA	LNP Plastics
Torlon	PAI	Amoco
Toyobo MXDA	PAMXD6	Toyobo
Trefsin	TPE/TPV	Exxon
Triax	PA/ABS alloys	Monsanto
Trithene	LDPE	Petroquimica Triunfo
Trithera	EVA	Petroquimica Triunfo
Trogamid T	amorphous PA	Huls
Tyril	SAN	Dow
Tyrin	CPE	Dow
Udel	PSU	Amoco Chemicals
Ugikral	ABS	Orkem
Ultem	PEI	GEP (General Electric Plastics)
Ultrablend	PBT/PC	BASF
Ultrablend S	PBT/ASA	BASF
Ultradur	PBT	BASF
Ultraform	POM-K	BASF
Ultralen	PET	BASF
Ultraplas	MF	Dynamit Nobel
Ultramid	PA 6	BASF
Ultramid A	PA 66	BASF
Ultramid C	PA copolymers	BASF
Ultramid S	PA 610	BASF
Ultranyl	PPO-M/PA or PPE/PA	BASF
Ultrason E	PES	BASF
Ultrason S	PSU	BASF
Urochem	UF	Chemiplastica Spa
Uroplast	UF	SterlingMoulding Materials
Urtal	ABS	Enichem
Valox	PBT	GEP (General Electric Plastics)
Vandar	Thermoplastic	Hoechst/Hoechst Celanese alloys

Trade name/ trade marks	Abbreviation	Supplier
Vectra	LCP	Hoechst/Hoechst Celanese
Vedril	PMMA	Enimont
Versamid	PA	
Verton	Long fiber filled thermoplastic	ICI
Vespel	PI	DuPont
Vestamid	PA 12	Huls
Vestoblend	PPE/PA	Huls
	PBT	Huls
Vestolen A	PE-HD	Huls
Vestolen P	PP	Huls
Vestolit	PVC	Huls
Vestopal	UP	Huls
Vestoplast	amorphous PO	Huls
Vestopren	EPM	Huls
Vestoran	PPO-M or PPE	Huls
Vestyron	PS	Huls
Vibrin	UP resins	Fiberglass
Viclan	PVDC/PVC	ICI
Victrex PEEK	PEEK	ICI
Victrex PES	PES	ICI
Vinoflex	PVC	BASF
Vipla	PVC	European Vinyl Corp.
Vista	PVC	Vista Chemicals
Vistalon	EPDM	Exxon
Vitalon	PA 46	Trade name used in Japan for Stanyl
Vitax	ASA	Hitachi Chemicals
Vydox	PTFE	DuPont
Vydyne	PA	Monsanto
Vydyne R	PA 66 (reinforced)	Monsanto
Vynite	PVC/NBR	Alpha Chemical & Plastics
Vythene	PVC/PU	Alpha Chemical & Plastics
Wacker Chemie	PE-HD	Wacker Polyathylen
Welite	PBT	Wellman Inc.
Wellamid	PA6 & PA66	Wellman Inc.
Welpet	PET	Wellman Inc.
Xantar	PC	DSM (Dutch State Mines)
Xenoy	PC/PBT	GEP (General Electric Plastics)
Xydar	LCP	Amoco
Zytel	PA 66	DuPont
Zytel ST	PA 66 super tough	DuPont

TABLE 5A. DENSITIES AND APPROXIMATE COSTS OF MATERIALS.

Full Name	Abbrev.	Density g/cc	Spec. Vol. cc/g	Cost p/g	Cost p/cc
Allyls:	**DAP**	1.80	0.56	0.82	1.47
	DAIP	1.80	0.56	1.275	2.28
Dough molding compounds: DMC					
General purpose grade		1.75	0.57	0.14	0.24
Arc and track resistant grade		1.73	0.58	0.15	0.25
Low profile grade		1.71	0.58	0.14	0.25
Epoxide molding compounds: EMC					
General purpose grade		1.90	0.53	0.26	0.50
Mineral glass filled		1.90	0.53	0.39	0.73
Long fibre grade		1.75	0.57	0.48	0.83
Granulated polyester molding compounds:	**GPMC**	2.00	0.50	0.13	0.26
Melamine Formaldehyde:	**MF**				
Compression grade		1.50	0.67	0.15	0.23
Injection grade		1.50	0.67	0.16	0.24
Melamine-Phenolics	**MPF**				
General purpose grade		1.60	0.63	0.15	0.24
Mineral filled grade		1.70	0.59	0.16	0.27
Phenol Formaldehyde:	**PF**				
Compression grade		1.40	0.71	0.00	
Injection grade		1.40	0.71	0.09	0.13
Sheet Molding Compounds: SMC					
General purpose grade		1.81	0.55	0.16	0.28
Low shrinkage (Profile) grade		1.82	0.55	0.17	0.30
Flame retardant - low shrinkage grade		1.74	0.57	0.20	0.35
Urea Formaldehyde:	**UF**				
Compression grade		1.50	0.67	0.09	0.14
Injection grade		1.50	0.67	1.00	0.15

N.B. The density values are typical but changes in density may occur due to the type and amount of filler used in a particular compound.

TABLE 5B. DENSITIES AND APPROXIMATE COSTS OF MATERIALS.

Full Name	Abbrev.	Density	Spec. Vol.	Cost	Cost
		oz/cu.in	cu.in/oz	c/cc	c/cu.in
Allyls:	**DAP**	1.04	0.96	37.04	38.55
	DAIP	1.04	0.96	57.73	57.80
Dough molding compounds:	**DMC**				
General purpose grade		1.01	0.99	6.25	6.29
Arc and track resistant grade		1.00	1.00	6.79	6.56
Low profile grade		0.99	1.01	6.25	6.56
Epoxide molding compounds:	**EMC**				
General purpose grade		1.10	0.91	11.77	13.12
Mineral/glass filled		1.10	0.91	17.66	19.15
Long fibre grade		1.01	0.99	21.73	21.78
Granulated polyester molding compounds:	**GPMC**	1.16	0.86	5.89	6.82
Melamine Formaldehyde:	**MF**				
Compression grade		0.87	1.15	6.79	6.03
Injection grade		0.87	1.15	7.24	6.29
Melamine-Phenolics	**MPF**				
General purpose grade		0.93	1.08	6.79	6.29
Mineral filled grade		0.98	1.02	7.24	7.08
Phenol Formaldehyde:	**PF**				
Compression grade		0.81	1.24	4.21	3.41
Injection grade		0.81	1.24	4.21	3.41
Sheet Molding Compounds:	**SMC**				
General purpose grade		1.05	0.95	7.02	7.37
Low shrinkage (Profile) grade		1.05	0.95	7.97	7.87
Flame retardant - low shrinkage grade		1.01	0.99	9.15	9.18
Urea Formaldehyde:	**UF**				
Compression grade		0.87	1.15	4.17	3.67
Injection grade		0.87	1.15	4.44	3.93

N.B. The density values are typical but changes in density may occur due to the type and amount of filler used in a particular compound.

As plastics materials are bought by weight and sold by volume (in the form of moldings) then, any material with a low density, or specific gravity (SG), has a tremendous advantage. Often therefore, the cost per unit volume is more important than the cost per unit weight. To convert to the cost in cents per cubic inch (c/cu in), multiply the cost per pound (lb) in

117

cents (c) by the SG and then, by 0.0361. Note that, the density in g/cc has the same numerical value as the SG.

To obtain density as oz/cu in, multiply g/cc by 0.5781.

To obtain density as lbs/cu in, multiply g/cc by 0.0361.

The conversions from one currency to another were obtained by assuming that there are 1.60 dollars to the pound sterling.

That is, $1.60 = £1.00 or 160 cents = 100 pence (p) or, 1.60c = 1p.

So, for example, to convert from p/g to c/g multiply the p/g by the currency conversion factor, for example, 1.60 in this case. To convert from p/g to c/oz multiply the p/g by the currency conversion factor and then by the number of grams in an ounce, that is, 28.3.

To convert from p/g to c/cu in: first convert to p/lb by multiplying by 454 (454g = 1lb), then to obtain c/lb multiply p/lb by, the currency conversion factor, for example, 1.60 in this c ase. To convert to the cost in cents per cubic inch (c/cu in), multiply the cost per pound (lb) in cents (c) by the SG and then, by 0.0361. That is:

$$p/g \times 454 \times 1.60 \times 0.0361 \times SG = c/cu\ in$$

$$p/g \times 26.22 \times SG = c/cu\ in$$

To convert from p/cc to c/cu, first multiply by the number of cubic centimeters (cc) in an ounce, that is, 16.39 and then by the currency conversion factor, for example, 1.60 in this case. So p/cc x 16.39 x 1.60 = c/cu in.

TABLE 6. APPROXIMATE PRICE OF MATERIALS RELATIVE TO GENERAL PURPOSE PHENOL FORMALDEHYDE (GPPF = 1.00).

Allyls:

DAP (General purpose and flame retardant grades)	8.8
DAIP (General purpose and flame retardant grades)	13.6

Dough molding compounds:

General purpose grade	1.5
Arc and track resistant grade	1.6
Low profile grade	1.6

Epoxy molding compounds:

Injection molding grade	2.8
Filled injection molding grade	4.2
Long fibre compression molding grade	5.1

Granulated polyester molding compound: 1.4

Melamine Formaldehyde:

Compression molding grade (white)	1.7
Compression molding grade (colors)	1.9
Injection molding grade (white)	1.7
Injection molding grade (colors)	1.9

Melamine Phenolic Formaldehyde:

Standard grades	1.6
Mineral filled grades	1.7

Phenol Formaldehyde:

General purpose grade	1.0
Electrical grade	
High impact grade	
Injection molding grade	

Sheet molding compounds:

General purpose grade	1.7
Low shrinkage (profile) grade	1.8
Flame retardant-low shrinkage grade	2.2

Urea Formaldehyde:

Compression molding grade (white)	1.0
Compression molding grade (colors)	1.2
Compression molding grade (special colors)	1.4
Injection molding grade (white)	1.1
Injection molding grade (colors)	1.3
Injection molding grade (special colors)	1.4

TABLE 7A. PROPERTY GUIDELINES.

Property	Units	ASTM	Allyl GF filled	Allyl Mineral filled	DMC
Tensile Strength	psi	D638	5,800-11,200	4,900-8,100	4,850-10,120
Elongation at Break	%	D638	3-5	3-5	1.5-2.0
Compressive Strength	psi	D695	24,700-35,200	20,200-31,750	14,300-30,300
Flexural Strength	psi	D790	8,800-19,500	8,700-10,850	16,300-24,000
Tensile modulus	10^3psi	D638	1,380-2,180	1,210-2,200	1,550-2,500
Flexural modulus	10^3psi	D790	1,250-1,520	1,100-1,375	1,100-3,050
Izod impact strength	'a'	D256A	0.4-14	0.3-0.75	4-12.5
Hardness–Rockwell		D785	E80-87	E61	–
Hardness–Barcol			–	–	40-70
Thermal expansion	'b'	D696	10-34	10-40	20-48
HDT 264psi	°C °F	D648 D648	165-290 329-654	160-290 320-554	160-180 320-356
Thermal conductivity	'c'	C177	5-15	7-25	18-22
Linear shrinkage	in/in	D955	0.0005-0.005	0.002-0.007	0.0005-0.004
Specific gravity		D792	1.7-1.98	1.68-1.86	1.7-2.07
Water abs in 24h	'd'%	D570	0.1-0.35	0.2-0.5	0.1-0.45
Dielectric str. – short time and with 0.125 samples	v/mil	D149	400-455	400-455	310-395

'a'   foot pounds per inch of notch: 0.125in specimens used
'b'   thermal expansion means the coefficient of thermal expansion with units of 10^{-6}in/in°C
HDT stands for heat deflection temperature or, deflection temperature under flexural load.
'c'   the units of thermal conductivity are 10^{-4}cal-cm/sec-cm^2 °C
'd'   water absorption tests performed with 0.125in specimens.

TABLE 7B. PROPERTY GUIDELINES.

Property	Units	ASTM	EMC GF Filled	EMC GF Filled High Str.	EMC Mineral Filled
Tensile Strength	psi	D638	4,800-19,500	17,500-27,000	3,700-10,200
Elongation at Break	%	D638	3-4	<3	<3
Compressive strength	psi	D695	17,500-39,000	29,000-37,500	17,000-40,000
Flexural Strength	psi	D790	8,000-28,000	45,500-67,000	6,000-18,000
Tensile modulus	10^3psi	D638	3,000-3,200	3,500-4,200	2,800-3,500
Flexural modulus	10^3psi	D790	1,800-4,500	4,500-8,000	1,400-2,100
Izod impact strength	'a'	D256A	0.3-0.7	25-34	0.3-0.5
Hardness–Rockwell		D785	M105-M110	M105-M110	– –
Hardness–Barcol			–	–	60-74
Thermal expansion	'b'	D696	10-52	–	20-60
HDT 264psi	°C	D648	93-271	149-288	93-260
	°F	D648	200-520	300-550	200-500
Thermal conductivity	'c'	C177	4-11	7-14	4-6
Linear shrinkage	in/in	D955	0.001-0.008	0.0002-0.0005	0.002-0.010
Specific gravity		D792	1.6-1.98	1.8-1.85	1.6-2.1
Water abs in 24h	'd'%	D570	0.04-0.2	0.02-0.2	0.02-0.2
Dielectric str. – short time and with 0.125 samples	v/mil	D149	240-380	350-400	240-390

'a'   foot pounds per inch of notch: 0.125in specimens used
'b'   thermal expansion means the coefficient of thermal expansion with units of 10^{-6}in/in°C
HDT   stands for heat deflection temperature or, deflection temperature under flexural load.
'c'   the units of thermal conductivity are 10^{-4}cal-cm/sec-cm^2 °C
'd'   water absorption tests performed with 0.125in specimens.

TABLE 7C. PROPERTY GUIDELINES.

Property	Units	ASTM	GPMC	MF Cellulose Filled	MF GF filled
Tensile Strength	psi	D638	3,000-8,500	4,500-14,000	4,700-10,200
Elongation at Break	%	D638	<3	0.6-1.0	<0.7
Compressive strength	psi	D695	12,500-37,000	32,000-44,500	18,000-35,000
Flexural Strength	psi	D790	5,500-16,000	8,500-44,000	13,000-23,000
Tensile modulus	10^3psi	D638	450-3,000	1,000-1,380	1,550-2,350
Flexural modulus	10^3psi	D790	1,800-2,000	900-1,100	900-1,000
Izod impact strength	'a'	D256A	0.3-0.5	0.2-0.4	0.6-15
Hardness–Rockwell		D785	E98	M110-M125	M110-M115
Thermal expansion	'b'	D696	18-48	40-45	15-25
HDT 264psi	°C °F	D648 D648	160-260 320-500	177-204 350-400	188-204 370-400
Thermal conductivity	'c'	C177	10-24	6-10	10-12
Linear shrinkage	in/in	D955	0.002-0.008	0.005-0.010	0.001-0.006
Specific gravity		D792	1.6-2.2	1.47-1.52	1.5-2.0
Water abs in 24h	'd'%	D570	0.04-0.45	0.1-0.5	0.1-1.2
Dielectric str. – short time and with 0.125 samples	v/mil	D149	240-380	250-400	120-370

'a' foot pounds per inch of notch: 0.125in specimens used
'b' thermal expansion means the coefficient of thermal expansion with units of 10^{-6}in/in°C
HDT stands for heat deflection temperature or, deflection temperature under flexural load.
'c' the units of thermal conductivity are 10^{-4}cal-cm/sec-cm^2 °C
'd' water absorption tests performed with 0.125in specimens.

TABLE 7D. PROPERTY GUIDELINES.

Property	Units	ASTM	MPF	PF Cotton Filled	PF Cellulose Filled
Tensile Strength	psi	D638	5,800-7,500	5,500-10,000	3,200-6,400
Elongation at Break	%	D638	<1	1-2	1-2
Compressive strength	psi	D695	25,000-30,000	22,000-30,500	22,000-30,000
Flexural Strength	psi	D790	8,000-9.500	8,500-12,000	6,000-11,000
Tensile modulus	10^3psi	D638	800-1,500	1,000-1,400	1,000-1,400
Flexural modulus	10^3psi	D790	1,000-1,200	700-1,250	850-1,250
Izod impact strength	'a'	D256A	0.1-0.4	0.2-1.0	0.3-1.1
Hardness–Rockwell		D785	E95-E100	M105-M115	M95-M115
Thermal expansion	'b'	D696	10-38	15-20	20-30
HDT 264psi	°C	D648	138-149	149-199	149-177
	°F	D648	280-300	300-390	300-350
Thermal conductivity	'c'	C177	4-7	8-10	6-8
Linear shrinkage	in/in	D955	0.009-0.011	0.004-0.010	0.004-0.010
Specific gravity		D792	1.5-1.7	1.36-1.43	1.36-1.43
Water abs in 24h	'd'%	D570	0.3-0.6	0.6-0.9	0.5-0.9
Dielectric str. – short time and with 0.125 samples	v/mil	D149	210-320	200-350	300-370

'a'   foot pounds per inch of notch: 0.125in specimens used
'b'   thermal expansion means the coefficient of thermal expansion with units of 10^{-6}in/in°C
HDT   stands for heat deflection temperature or, deflection temperature under flexural load.
'c'   the units of thermal conductivity are 10^{-4}cal-cm/sec-cm^2 °C
'd'   water absorption tests performed with 0.125in specimens.

TABLE 7E. PROPERTY GUIDELINES.

Property	Units	ASTM	PF Woodflour Filled	PF GF Filled High Str.	PF Mineral Filled
Tensile Strength	psi	D638	4,800-8,500	6,500-17,000	5,700-8,500
Elongation at Break	%	D638	0.4-0.8	<0.3	0.1-0.5
Compressive strength	psi	D695	23,500-30,000	25,000-65,500	22,000-34,000
Flexural Strength	psi	D790	7,000-12,000	15,500-55,500	10,000-13,000
Tensile modulus	10^3psi	D638	700-1,500	1,800-3,000	2,000-2,300
Flexural modulus	10^3psi	D790	1,000-1,250	2,500-25,000	1,000-1,800
Izod impact strength	'a'	D256A	0.2-0.5	0.5-15	0.2-0.35
Hardness–Rockwell		D785	M100-M115	E50-E100	E80-E90
Thermal expansion	'b'	D696	30-45	5-20	18-25
HDT 264psi	°C	D648	149-177	177-288	177-232
	°F	D648	300-350	350-550	350-450
Thermal conductivity	'c'	C177	4-8	7-14	20-25
Linear shrinkage	in/in	D955	0.004-0.010	0.001-0.004	0.002-0.007
Specific gravity		D792	1.36-1.45	1.7-2.0	1.42-1.84
Water abs in 24h	'd'%	D570	0.3-1.2	0.04-1.2	0.1-0.3
Dielectric str. – short time and with 0.125 samples	v/mil	D149	240-380	150-400	200-340

'a' foot pounds per inch of notch: 0.125in specimens used
'b' thermal expansion means the coefficient of thermal expansion with units of 10^{-6}in/in°C
HDT stands for heat deflection temperature or, deflection temperature under flexural load.
'c' the units of thermal conductivity are 10^{-4}cal-cm/sec-cm^2 °C
'd' water absorption tests performed with 0.125in specimens.

TABLE 7F. PROPERTY GUIDELINES.

Property	Units	ASTM	SMC	SMC Low Shrink	UF Cellulose Filled
Tensile Strength	psi	D638	7,500-24,500	4,000-19,500	5,000-12,500
Elongation at Break	%	D638	2.5-3	3-4.5	<1
Compressive strength	psi	D695	14,500-30,000	14,500-30,000	22,000-40,000
Flexural Strength	psi	D790	9,000-35,000	8,500-35,500	10,000-16,000
Tensile modulus	10^3psi	D638	1,400-2,400	1,000-2,400	1,000-1,400
Flexural modulus	10^3psi	D790	1,000-2,100	1,000-2,400	1,200-1,500
Izod impact strength	'a'	D256A	6.5-20	1.5-15	0.2-0.35
Hardness–Rockwell		D785	–	–	M110-M120
Hardness–Barcol		D785	50-70	40-70	–
Thermal expansion	'b'	D696	18-20	5-20	20-35
HDT 264psi	°C	D648	188-260	188-260	121-143
	°F	D648	370-500	370-500	250-290
Thermal conductivity	'c'	C177	15-21	15-21	7-10
Linear shrinkage	in/in	D955	0.001-0.004	0.0001-0.002	0.006-0.010
Specific gravity		D792	1.6-2.5	1.6-2.4	1.45-1.52
Water abs in 24h	'd'%	D570	0.1-0.25	0.01-0.25	0.4-0.8
Dielectric str. – short time and with 0.125 samples	v/mil	D149	350-500	350-500	300-400

'a'   foot pounds per inch of notch: 0.125in specimens used

'b'   thermal expansion means the coefficient of thermal expansion with units of 10^{-6}in/in°C

HDT   stands for heat deflection temperature or, deflection temperature under flexural load.

'c'   the units of thermal conductivity are 10^{-4}cal-cm/sec-cm^2 °C

'd'   water absorption tests performed with 0.125in specimens.

TABLE 8. UNIT CONVERSION

Length

Knowing	GAUGE	multiply by 0.254	to get	MICRONS
Knowing	THOU	multiply by 25.4	to get	MICRONS
Knowing	INCHES	multiply by 25.4	to get	MILLIMETERS
Knowing	FEET	multiply by 0.305	to get	METERS
Knowing	YARDS	multiply by 0.914	to get	METERS
Knowing	MICRONS	multiply by 3.937	to get	GAUGE
Knowing	MM	multiply by 0.0394	to get	INCHES
Knowing	CM	multiply by 0.3937	to get	INCHES
Knowing	METERS	multiply by 1.094	to get	YARDS
Knowing	METERS	multiply by 3.28	to get	FEET

Area

Knowing	SQ. INCHES	multiply by 645	to get	SQ. MM
Knowing	SQ. FEET	multiply by 0.093	to get	SQ. METERS
Knowing	SQ. YARDS	multiply by 0.8354	to get	SQ. METERS
Knowing	SQ. MM	multiply by 0.0016	to get	SQ. INCHES
Knowing	SQ. METERS	multiply by 10.76	to get	SQ. FEET

Volume

Knowing	CUBIC INCHES	multiply by 16,387	to get	CUBIC MILLIMETERS
Knowing	CUBIC INCHES	multiply by 16.39	to get	CUBIC CENTIMETERS
Knowing	CUBIC FEET	multiply by 0.0284	to get	CUBIC METERS
Knowing	CUBIC CENTIMETERS	multiply by 0.061	to get	CUBIC INCHES
Knowing	CUBIC METERS	multiply by 35.3	to get	CUBIC FEET
Knowing	US LIQUID PINTS	multiply by 0.833	to get	BRITISH IMP. PINTS
Knowing	US LIQUID PINTS	multiply by 0.473	to get	LITERS
Knowing	BRITISH IMP. PINTS	multiply by 1.201	to get	US LIQUID PINTS
Knowing	LITERS	multiply by 2.114	to get	US LIQUID PINTS
Knowing	LITERS	multiply by 61.024	to get	CUBIC INCHES

Weight

Knowing	OUNCES	multiply by 28.3	to get	GRAMS
Knowing	GRAMS	multiply by 0.0353	to get	OUNCES
Knowing	POUNDS	multiply by 0.452	to get	KILOGRAMS
Knowing	TONS	multiply by 0.907	to get	METRIC TONNES
Knowing	KILOGRAMS	multiply by 2.205	to get	POUNDS
Knowing	METRIC TONNES	multiply by 1.103	to get	TONS

Pressure and Stress

Knowing PSI	multiply by 0.0069	to get	MPa
Knowing MPa	multiply by 145	to get	PSI
Knowing MPa/MNm^{-2}	multiply by 10	to get	BAR
Knowing BAR	multiply by 1.013	to get	ATMOSPHERES
Knowing BAR	multiply by 14.50	to get	PSI

Speed

Knowing FEET/ MINUTE	multiply by 0.0051	to get	METERS/ SECOND
Knowing FEET/ MINUTE	multiply by 0.305	to get	METERS/ MINUTE
Knowing METERS/ SECOND	multiply by 197	to get	FEET/MINUTE
Knowing METERS/	multiply by 3.28	to get	FEET/MINUTE

Density

Knowing POUNDS/ CUBIC FOOT	multiply by 16	to get	KILOGRAM/ CUBIC METER
Knowing KILOGRAMS/ CUBIC METER	multiply by 0.06243	to get	POUNDS/ CUBIC FOOT
Knowing GRAMS/ CUBIC CENTIMETRE	multiply by 0.58	to get	OUNCES/CUBIC INCH
Knowing OUNCES/ CUBIC INCH	multiply by 1.73	to get	GRAMS/CUBIC CENTIMETER.
Knowing SPECIFIC GRAVITY	multiply by 62.4	to get	POUNDS/CUBIC FOOT
Knowing SPECIFIC GRAVITY	multiply by 16.39	to get	GRAMS/CUBIC INCH
Knowing SPECIFIC GRAVITY	multiply by 0.58	to get	OUNCES/CUBIC INCH
Knowing SPECIFIC GRAVITY	multiply by 0.036	to get	POUNDS/CUBIC INCH

Energy, Quantity of Heat and Work

Knowing FOOT POUNDS	multiply by 1.356	to get	JOULES
Knowing BTUS	multiply by 1.056	to get	KILOJOULES
Knowing BTUS	multiply by 778.2	to get	FOOT POUNDS
Knowing BTUS	multiply by 0.252	to get	KILOCALORIES
Knowing BTUS/H	multiply by 0.2931	to get	WATTS
Knowing BTUS	multiply by 2.928 x 10^{-4}	to get	FOOT POUNDS
Knowing JOULES	multiply by 0.738	to get	FOOT POUNDS
Knowing KILOJOULES	multiply by 0.9478	to get	BTUS
Knowing KILOJOULES	multiply by 1000	to get	JOULES
Knowing CALORIES	multiply by 4.184	to get	JOULES

Power

Knowing	FOOT POUNDS/SECOND	multiply by 1.356	to get	WATTS
Knowing	HORSE-POWER	multiply by 745.7	to get	WATTS
Knowing	HORSE-POWER	multiply by 550	to get	FOOT POUNDS/SECOND
Knowing	KILOWATTS	multiply by 737.6	to get	FOOT POUNDS/SECOND
Knowing	KILOWATTS	multiply by 1.341	to get	HORSEPOWER

TEMPERATURE CONVERSION

In the temperature conversion table (see table 10) the center column contains the numbers that you wish to convert. If you have a temperature in °C that you would like in °F then, read from the center to the right. For example, if you would like to convert 200°C into °F, then find 200 in the center column and then look immediately right: you should see 392°F. Check this by dividing by 5, multiplying by 9 and adding on 32.

$$\frac{200 \times 9}{5} = 360.$$

Then 360 + 32 = 392. (or 200 x 1.8 then +32).

To go from °F to °C then enter the figure you have in the center column and look immediately left. 200°F becomes 93°C. Check this by taking away 32 from 200, dividing by 9 and then multiplying by 5.

200 − 32 = 168.

Then $\frac{168 \times 5}{9} = 93$. (or 168 x 0.56)

TABLE 10. TEMPERATURE CONVERSION TABLE.

Centigrade	Starting Value	Fahrenheit
−18	0	32
−12	10	50
−7	20	68
−1	30	86
4	40	104
10	50	122
16	60	140
21	70	158
27	80	176
32	90	194
38	100	212
43	110	230
49	120	248
54	130	266
60	140	284
66	150	302
71	160	320
77	170	338
82	180	356
88	190	374
93	200	392
99	210	410
104	220	428

Centigrade	Starting Value	Fahrenheit
110	**230**	446
116	**240**	464
121	**250**	482
127	**260**	500
132	**270**	518
138	**280**	536
143	**290**	554
149	**300**	572
154	**310**	590
160	**320**	608
166	**330**	626
171	**340**	644
177	**350**	662
182	**360**	680
188	**370**	698
193	**380**	716
199	**390**	734
204	**400**	752
210	**410**	770
216	**420**	788
221	**430**	806
227	**440**	824
232	**450**	842
238	**460**	860
243	**470**	878
249	**480**	896
254	**490**	914
260	**500**	932
266	**510**	950
271	**520**	968
277	**530**	986
282	**540**	1004
288	**550**	1022
293	**560**	1040
299	**570**	1058
304	**580**	1076
310	**590**	1094
316	**600**	1112
321	**610**	1130

Centigrade	Starting Value	Fahrenheit
327	**620**	1148
332	**630**	1166
338	**640**	1184
343	**650**	1202
349	**660**	1220
354	**670**	1238
360	**680**	1256
366	**690**	1274
371	**700**	1292
377	**710**	1310
382	**720**	1328
388	**730**	1346
393	**740**	1364
399	**750**	1382
404	**760**	1400
410	**770**	1418
416	**780**	1436
421	**790**	1454